AF710322

Ecological Studies

Analysis and Synthesis

Edited by

W. D. Billings, Durham (USA) F. Golley, Athens (USA)

O. L. Lange, Würzburg (FRG) J. S. Olson, Oak Ridge (USA)

Volume 23

Francis Dov Por

Lessepsian Migration

The Influx of Red Sea Biota into the Mediterranean by Way of the Suez Canal

With 47 Figures, 10 Plates, and 2 Maps

Springer-Verlag Berlin Heidelberg New York 1978

Professor F. D. POR
Department of Zoology
The Hebrew University of Jerusalem
Jerusalem/Israel

QH
195
.E4
P67

Cover motive based on NASA satellite picture taken from Gemini XI.

ISBN 3-540-08381-2 Springer-Verlag Berlin Heidelberg New York
ISBN 0-387-08381-2 Springer-Verlag New York Heidelberg Berlin

Library of Congress Cataloging in Publication Data. Por, Francis Dov, 1927-. Lessepsian migration. (Ecological studies; v. 23). Bibliography: p. Includes index. 1. Marine ecology—Egypt—Suez Canal. 2. Marine ecology—Mediterranean Sea. 3. Marine ecology—Red Sea. 4. Animal migration. I. Title. II. Series. QH 195. E 4 P 67.574.5′2.77-24546.

This work is subject to copyright. All rights are reserved, whether the whole or part of the material is concerned, specifically those of translation, reprinting, re-use of illustrations, broadcasting, reproduction by photocopying machine or similar means, and storage in data banks.
Under § 54 of the German Copyright Law where copies are made for other than private use, a fee is payable to the publisher, the amount of the fee to be determined by agreement with the publisher.

© by Springer-Verlag Berlin · Heidelberg 1978.
Printed in Germany.

The use of registered names, trademarks, etc. in this publication does not imply, even in the absence of a specific statement, that such names are exempt from the relevant protective laws and regulations and therefore free for general use.

Typesetting, printing, and binding: Brühlsche Universitätsdruckerei, Lahn-Gießen.
2131/3130—543210

To the Memory of my Father

Preface

There are few achievements of modern man which can compare to the Suez Canal. In Egypt—the land of the most famous wonders of antiquity—the Suez Canal was built as the first technical wonder of the industrial revolution. Ferdinand de Lesseps was a man straddling two epochs—the romantic utopism of Saint-Simon and the modern world of technocracy. The gigantic project was at its start shouldered by the crowds of tens of thousands of forced laborers still available and ended as a show-piece for modern mechanical earth-moving techniques.

The canal builders themselves were still polyhistors in the old sense: engineers-cum-zoologists; naval officers-cum-geologists; diplomats meddling with chemistry. During the four generations of the existence of the Suez Canal, the fateful professional narrowmindedness became progressively worse. The engineers continued their work in and around the Canal, but they became oblivious and unresponsive to the environmental impact, to the fascinating changes in the biotic scenery which they were producing with their own hands.

The Suez Canal started as an international enterprise, a center of preoccupation for the most enlightened minds of the mid 19th century. The same was true for the nascent study of marine biology. During the century which has since passed, however, the international commonwealth of scientific research has fallen apart, and the Suez Canal has become an increasingly coveted object in the international power struggle. The more the cannons roared, the more the muses of science fell silent. Once the center of the intellectual world, the Suez Canal has disappeared from the map of modern science.

The biggest scientific enterprise carried out in the Suez Canal in this century was the Cambridge Expedition—a short research trip (some 2 months) of three young English scientists headed by H. Munro Fox (November–December 1924). The situation changed somewhat with the work of the first generation of local marine biologists—H. A. F. Gohar in Egypt and W. and H. Steinitz (father and son) in Israel.

For an ephemerous moment in 1951, Gohar, Steinitz and two other promoters of Suez Canal research, Kosswig and Tortonese, met in Istanbul under the auspices of UNESCO. Fifteen years later UNESCO tried to sponsor a regional meeting on the impact of the Suez Canal migration on the Mediterranean biota. The projected conference changed conveners and proposed meeting place from Cyprus to Dubrovnik until, in 1970, the plan was finally abandoned.

In 1966, at the suggestion of Professor G. Thorson, Professor H. Munro-Fox asked Professor H. Steinitz and myself to write a review on migration through the Suez Canal. It is with a sense of veneration towards my recently deceased teachers and colleagues that I have now carried out this project. It is also as a token of admiration to Ferdinand de Lesseps that I coined the term "Lessepsian migration" for the outstandingly fascinating phenomenon of the invasion of the Mediterranean sea by the tropical Red Sea fauna and flora.

This is a book for marine biologists. It is limited to the topic of Lessepsian migration itself, without attempting either to discuss the Suez Canal as an aquatic environment or to analyze the dynamic changes which have occurred in the Levant Basin of the Mediterranean. This would have been an impossible task, since the subject is still badly in need of advanced interdisciplinary methods of study, and is far beyond the scope of a one–man enterprise.

This study is the fruit of my involvement with this subject over many years. It is an effort to explain and forecast and is the expression of a humanistic approach in which geological and human history interweave with basic biogeography and a modern environmentalist approach.

Many of the shortcomings of my single-handed efforts have been overcome thanks to the help of my colleagues. I am especially grateful for the critical reading, typing, and checking of Miss I. Ferber; for the opportunity given to me by the late Professor H. Steinitz and by Dr. W. Aron to extract the quintessence of seven years of research effort made by the team of the Hebrew University–Smithsonian Institution Joint Program, *Biota of the Red Sea and Eastern Mediterranean;* for the many pieces of information and advice amicably shared with me by, among others, Professor Al. Barash, Dr. R. H. Benson, Professor A. Ben-Tuvia, Professor B. Kimor, Dr. Y. Lipkin, Mr. O. H. Oren and Dr. I. Paperna; and finally, for the manifold cooperation of scores of students and assistants who, by the very process of their learning, taught me too. Support from the Israel Academy of Sciences and Humanities is gratefully acknowledged.

It is hoped that this book, a summary of the current thinking on this unique migrational process, will serve as a stepping stone for further research.

Jerusalem, July 1978 F. D. Por

Contents

Conspectus

1. "Lessepsian Migration"

The reader should first be introduced to the term "Lessepsian migration", which is as yet of only limited circulation. The term was coined (Por, 1969a, 1971b) when it became evident that the unidirectional migration of the Red Sea species to the Mediterranean via the Suez Canal is a unique phenomenon in modern biogeography. This uniqueness in the dimensions and directionality of the migration through the Suez Canal needed a short and diagnostic name. As such, "Lessepsian migration" is a phenomenon of unidirectional and successful biotic advance from the Red Sea to the Eastern Mediterranean. It does not include migrational happenings which occurred through passive transport to harbors far from the Canal; nor does it include the rather isolated cases of migration in the opposite sense: from the Mediterranean to the Red Sea. Finally, it does not include the many cases of migration into the Canal, though many of such species might be future Lessepsian migrants.

2. The Structure of the Book

Even with the limitations of the term "Lessepsian migration" as given above, the subject has to be treated in a fairly wide perspective: some biogeographic aspects have to be clarified and the geological history of the faunal divide between the Mediterranean and the Indian Ocean has to be dealt with. Not only the marine coastal environments and the Suez Canal itself, but also the Levant Basin of the Mediterranean must be presented as the stage upon which the migrational happening is taking place; this includes some lagoon environments which serve as terms of comparison.

The book is subdivided into three main sections: the first deals with the geological–historical background, the second with the Suez Canal itself, and the third with the Lessepsian migrants in their new environment in the Mediterranean. Two final chapters deal with the cases of contrary migration ("anti-Lessepsian migration"), and a comparison with other inter-oceanic canals and biotic exchanges.

3. A Preamble to the Conclusions

The structure of the book as exposed above takes a somewhat meandering course and might therefore impose some difficulty on the less patient reader who is interested chiefly in grasping the conclusions of the work. We therefore give here a brief summary of the conclusions.

1. The separation of the Atlantic biota from the Indo-Pacific biota is as old as the late Miocene, and the significance of Lessepsian migration as a contact established after some ten millions of years is accordingly great.

2. The key to an understanding of Lessepsian migration lies in the analysis of the Pleistocene adaptations of the Red Sea and Mediterranean biota, the environmental conditions of the Suez Canal, primarily those of the Bitter Lakes and Lake Timsah, and also the diversity of the Eastern Mediterranean biota as a function of the present environmental conditions of this sea.

3. The Suez Canal is, generally speaking, a metahaline-marine environment. Its biota has reached, during its hundred years' existence, a certain successional maturity with a high level of diversity. The Lessepsian migrants are, as a rule, recruited from among the Suez Canal biota.

4. The Lessepsian migrants, assumed to number around 500 species, of which over 200 are listed, are presently concentrated along the Levant Coasts. Fishes and decapod crustaceans, as well as a few molluscs, have advanced beyond the limits of the Levant Basin.

5. There are no signs of invasional population patterns (i.e. rapid build-up and then population crash), but rather of a gradual increase and advance as against an accommodation or withdrawal of local competing species. No evolutionary changes of discernible dimensions are to be found as yet among the new settlers.

6. There is no indication that Lessepsian migration is speeding up. It seems, rather, that the process tends to an asymptotic equilibrium. The effects of the cessation of the Nile flow are as yet uncertain. In the present geoclimatic episode the Lessepsian migrants are limited to the Levant Basin and the North African gulfs.

7. Lessepsian migration is more similar to the postglacial faunal interchanges between the mediterranean sill-limited seas than to the faunal movements (or lack of movements) through other existing interoceanic shipping canals.

The reader who expects to find an analysis of Lessepsian migration along the lines of modern quantitative ecology will be disappointed. The dimensions of the phenomenon and the lack of uniformity of the available data are such as to make attempts of this kind impossible. Applying new research methods to Lessepsian migration will no doubt result in new insights in the future. But years will pass before sufficient specific cases from different taxa have been investigated; meanwhile, the existing data should be fully exploited to provide some working hypotheses for a process which commenced just over one century ago.

Dynamic Zoogeography

Zoogeography, like paleontology and evolution research is a historical-descriptive discipline. The conclusions are circumstantial and largely derived from comparative reasoning. In very few cases, and only at a microlevel, can the real dynamism be observed and hints be obtained of the deterministic network of the major, historical macrophenomena. In the modern man-molded ecosystems some evolutionary processes became accelerated and thus observable. However, without being entirely artificial epiphenomena, the human-influenced happenings are to a large extent new events in an evolving biosphere. Animals are selected naturally as well as artificially to coincide with the presence and needs of the human species. Animals preadapted or adapted to adventive spread as "camp followers", "hitch-hikers", etc., are the only ones which take advantage of human mobility. Lindroth (1956) draws a convincing picture of the set of adaptations which enable insects to cross the Atlantic from Europe to America as adventives on ships. A similar picture could probably also be drawn for the ecological valence which enables a marine organism to become an adventive diaspore on a ship's hull. To my knowledge, no listing of these qualities has ever been made. A fouling-cosmopolite is an animal with a sessile life, a nonspecific filter feeder, very adaptative in its reproduction requirements, euryhaline, resistant to current, and resistant to pollutants.

All these characteristics ipso facto also guarantee the spread of the ship-carried pioneers in their new environment, and ensure their competitive success against the local fauna.

As with other shipping lanes, the Suez Canal also served to spread the adventive marine fauna. There is no doubt that some of the well-known cosmopolitan fouling organisms of eastern origin such as *Mercierella enigmatica* (Polychaeta)[1] and *Elminius modestus* (Cirripedia) took advantage of the short sailing time through the Suez Canal and thus succeeded in their advance. It has however, by no means been proved that the Suez Canal was the only or necessary way to spread them into the Western hemisphere.

In the following treatment of Lessepsian migration I shall not consider the cosmopolitan fouling organisms. Their spread has nothing to do with the specific problem of Lessepsian migration. The emphasis will lie on the natural dynamism of the spreading phenomena, some of which include, however, some passive dispersal.

[1] Zibrowius (1973) recently questioned the alleged worldwide distribution of this species.

In zoogeography, as well as evolution, there is a continuous flow of changes, but in the short human perspective only a small time-glimpse of the major unfolding can be witnessed. Zoogeography is in a better situation than some of the related historico-evolutionary sciences: we are the witnesses of a major unsettling of the animal distribution owing to the Pleistocene glaciations. In the present period we are living through a warm interglacial with fauna gradually spreading back over the ice- and frost-damaged areas. Lindroth (1956) has described many cases of northward spread of animals into presently ice-free Scandinavia. According to Udvardy (1969, p. 150): "Temperate and subpolar ecosystems were depleted by the recent, drastic glaciations, and have only recently begun their renewed evolution". This statement is true not only for the terrestrial fauna; a depletion and destruction of biota also occurred in the sea.

Whereas in the open oceans the Pleistocene fluctuations caused mere advances and retreats, narrowing or widening of area, the situation was different in the inner, mediterranean basins such as the Mediterranean and Red Seas. Here the changes had zoogeographically defined dimensions: whole series of species disappeared from a given marine basin and at present are gradually returning.

Many data on this fluctuating dynamism of the Pleistocene mediterranean biota can be found in Segerstrale (1957), Mars (1963), Pérès (1967), Pusanov (1967), and Por (1975a). These are slow decennial advances of littoral species from the straits connecting with the open ocean inward into the mediterranean. For fish or plankton these are spotlike appearances of reproducing or metamorphosing swarms within sterile or larval populations that enter the sea (Fig. 1).

In the mediterranean seas, depleted by Pleistocene temperature and salinity fluctuations, the resettling process is perhaps even more marked than in the terrestrial habitats. Repopulation proceeds only through the narrow Gibraltar or Bosphorus Straits and the gradient of advance is very near to linear. On the other hand, the far end of the mediterranean is a hermetically closed depauperation area, a cul de sac comparable only to such areas as the southern tip of South America (Darlington, 1959). In these marine cul de sacs—like the farthest Eastern Mediterranean or the Gulf of Bothnia—the influence of the most extremely expressed environmental fluctuation is complemented by a "distance and time barrier" (Udvardy, 1969). The period of the Pleistocene fluctuation was in this case too short for the decennially spreading species to reach the far end of the Mediterranean.

The fact that the Eastern Mediterranean and the Gulf of Suez of the Red Sea are cul de sacs of two mediterranean seas had, and still has a decisive influence on the quality and direction of the faunal interchange through the Suez Canal. This has already been emphasized by Ben-Tuvia (1966) and Por (1971b) and will recurrently be discussed in the following chapters.

The above-mentioned postglacial (or interglacial) faunal adjustments are small-step changes in which isolated species only are involved. Following Dansereau (1957) and Bănărescu and Boşcaiu (1973), the zoogeography should study the repartition and dynamics of whole communities and ecosystems, of "syntaxons" in their words. De Lattin (1967) also discussed Pleistocene movements in terms of relative, often antagonistic, movements of three basic terrestrial types of ecosystems: the arboreal, the eremial and the oreo-tundral. These movements can be

Fig. 1. Alternative pulsations of glacial versus interglacial faunal elements in the mediterranean seas (original)

compared to the major faunal replacements of evolutionary history which, as a kind of "mesoevolution", form the connecting link between the small genetic adjustments at the species level and the large-scale typogenesis. There is no chance for a student of animal evolution to witness such a major faunal replacement in our "noosophere", the human-dominated biosphere.

In the field of zoogeography such chances are also reduced. However, some catastrophic or sudden events of major zoogeographic dimensions and/or dynamic speed could be studied.

Perhaps the best-known case was that of the Krakatau explosion. As summed up by Dammerman (1948), the Indonesian island was entirely depopulated by a volcanic eruption in 1883. Since then repopulation has occurred with increasing speed, giving precious insights into the spreading and pioneering capacities of the terrestrial animals, the mechanism of colonization and establishment of ecosystemic interrelations. It is necessary to mention here that zoologists and botanists have not taken advantage of the opportunities offered by the Krakatau catastrophe. Despite the fact that the surrounding fauna was well-known and that perfect experimental conditions existed in the form of the total barreness of Krakatau, the repopulation process was not followed in detail. The distances involved, the expeditionary conditions, the necessity of a coordinated research work by many scientists, and, not least, the war and the political situation, are responsible for the partial loss of a great research opportunity.

Krakatau-type catastrophes and repopulations have occurred repeatedly in geological history. Perhaps the most quoted case of a considerable geographic dimension was the lava extrusion of the Deccan traps in India (de Beaufort, 1951). Life was destroyed over a vast stretch of the Indian peninsula which consequently had to be repopulated from outside.

Generally, however, such extreme annihilations are exceptions to the majority of zoogeographic redistributions. A distributional province, as a rule, supports a set of ecosystems and is separated by a barrier from another province occupied by a different set of ecosystems. If the barrier is nearly hermetic the zoogeographic pattern is stable, but as soon as the barrier breaks down, the ecosystems of the two previously separated provinces interpenetrate. The result is a major distributional happening which changes the zoogeographical pattern of a given part of the globe. Barriers break down as a result of geological changes in the earth's crust, or major climatic or oceanic changes which modern mankind has no chance to witness. The artificial—or rather semi-artificial—opening of the Isthmus of Suez is the only example of a barrier breakdown which can be studied.

There is a gradation in the degree of tightness of the zoogeographic barriers. According to Udvardy (1969, p.18), the topographic, physical barriers "affect most animals by their ecological unfitness to support the dispersing pioneers". This is the most impenetrable type of barrier and to this category belong the wide stretches of sea serving as barriers for the terrestrial faunas and the continental masses or isthmi serving as barriers for the marine faunas. In this sense, the Isthmus of Suez was perhaps one of the most hermetic barriers of the biosphere, despite its narrowness of only 162 km. The breakdown of a tight barrier rarely leads to a completely unhindered intermixing of the faunas on both sides of the old obstacle. This is especially the case when the barrier lies across a longitudinal-radial pathway. The resulting "fusion faunas" are in Mayr's words (1965b) "particularly interesting as suitable material for the testing of zoogeographic methods".

When a land bridge through the emerging Central American Isthmus was established between the two Americas, the faunal interchange was not unhindered. Using Simpson's (1965) household word, a "filtrating bridge" resulted, which as a paraphrase to Udvardy (1969) is ecologically able to support only some of the dispersing pioneers. The Suez Canal is an aquatic replica of the filtrating bridge: a "sieving funnel" through which certain taxa can pass whereas others are retained.

The waters of the Suez Canal are not only a pathway through which dispersing pioneers are funneled, but, as convincingly shown by Steinitz (1968), they are also a "habitat", an ecosystem in their own right. Por (1973c) has shown that the biotic assemblage of the Bitter Lakes—the main environment of the Suez Canal—has progressed through typical successional stages and reached a climax of its own. This is a good example of what Udvardy (1969) calls a "living environment barrier". In other words, the biotic associations themselves—being more or less penetrable to competing immigrants—may serve as additional plugs in a funnel. For Udvardy, evidence of this type of barrier is circumstantial. In the case of the Suez Canal, its role can be better substantiated.

The direction of the movement through the zoogeographic pathways of the filtrating bridge is another problem. The interchange between two provinces is always unequal: the influence of one province on the other is stronger; one taxon might spread better in one sense, another may be more successful in another. In the well-known case of the two Americas the mammals of North America spread with numerous species over almost all of South America, whereas the mammals of South America had only a numerically and geographically limited success in North America. The result was a predominant invasion of the Nearctic fauna into the Neotropic region. Darlington (1959) explains these one-sided invasions in general terms as the result of a competitive advantage of the dominant biota which evolved in the large zoogeographic unit (with warmer and more stable conditions) over those of the small one (in this case the Darlingtonian Megagea).

In the case of the Suez Canal, the donor of almost all of the migrating species is the tropical Indo-West-Pacific region—the biggest and qualitatively richest of all the marine zoogeographical regions.

Following the work of Andrewartha and Birch (1954), it is well accepted that every species has an "innate tendency towards dispersal". Leston (1957) reiterates this notion and speaks of a specific "spreading potential" of every animal family.

The success of a migrational advance is not determined only by ability to jump the hurdles of the filtrating bridge but also by the ability to find a niche in the new ecosystem, and to maintain it through successful reproduction. Many factors and components are covered by the term "niche", but its accepted meaning being an interspecific web of relations, even if it is indirectly, though no less fatefully, determined by the abiotic ecologic valences of the competing species.

In the case of the Lessepsian migration—unlike the example of Krakatau—the new environment was not virgin and empty but inhabited by an ecosystem with a certain amount of environmental resistance and a limited selection of empty niches. The phenomenon is further characterized by the fact that not only a few hardy migrants crossed from the Red Sea to the Mediterranean, but hundreds of species, frequently perhaps whole sections of an original food web; so that nothing like the Eltonian explosive populations of invaders could be witnessed.

Since the separation of the Mediterranean from the Indo-West-Pacific is as old as the Miocene period, there were extremely few species pairs present on both sides of the Isthmus of Suez. This is a basic difference from the case of the geologically much younger Central American Isthmus. Because of the lack of congeneric pairs, there was and probably is no direct competition between aborigines and invaders. Following the more recent statements of Schminke (1973), congenerity means similar feeding mechanism and requirements. In the case of Lessepsian migration, to the extent to which the presently recognized taxonomic genera in different families are reliable, one cannot expect many instances of direct competition for the same position in the food web.

The genetically isolated emigrant diasporas, living under different abiotic and biotic surroundings, have most probably already started on the way to speciation.

Therefore, the Lessepsian migration may serve not only as an example of the establishment of a new zoogeographic province, but also of the evolutionary consequence of this—the appearance of vicariant (i.e. zoogeographically replacing) species which can be monitored.

The Lessepsian migrants are near-ideal cases for what Mayr (1965a) calls "peripheral isolates". These isolates, connected to their parental populations only through the gene trickle via the narrow Suez Canal, are prone to undergo major ecologic shifts leading to speciation.

1. The Historical Background

1.1 The Tethys and the Origins of the South West Asiatic Barrier

In order to appreciate the subject of the present monograph, one needs an understanding of the historical perspective back to the time of the Tethys Sea. This might seem purely academic, but the worldwide circumtropical marine connection which ceased to exist about 12 million years ago is the methodological baseline for the marine zoogeographical processes in our area.

The very existence of the question is based on two fundamental misunderstandings: first, there never was a homogenous Tethys fauna at the species level which extended all around the tropical belt. Even in the "golden age" of the Tethys, there were well-established provinces with vicariant species within this belt. The second misunderstanding is related to the antiquity of the breakup of the Tethys and the extreme environmental changes which followed: there is absolutely no possibility of any conspecific populations surviving in situ on both sides of the faunal divide of South West Asia.

Nevertheless, the existence of "Tethys relicts" has time and again been accepted by different authors: Kosswig (1942), Tortonese (1951), Steinitz (1951), Ekman (1967), Klausewitz (1968), Kosswig (1967) and de Lattin (1967). Upon closer scrutiny of the articles by these authors, one finds that there are actually no examples of conspecific relicts given, and so-called relicts are merely representatives of tropical genera still found today in the Mediterranean.

Ekman (1967, p. 88) states carefully: "As regards the Mediterranean, the term ("Tethys relicts") should imply that the organisms in question owe their presence in this sea solely to the former extent of the Tethys Sea in the Mediterranean and that they cannot have died out there ... and afterwards migrated into it". Nevertheless, in the same paragraph it follows: "... The number of real Tethys relicts may, however, be not inconsiderable".

Kosswig (1956, p. 86) states, concerning the Mediteranean: "There are grounds for supposing that the ancient tethic fauna has entirely disappeared ...". However, the same author (1967, p. 322) admits the possibility that the fish *Aphanius dispar* ... "left colonies behind it in Palestine during its retreat towards the Indian Ocean before the closing of the Mediterranean in the East".

Let us now discuss the facts. A marine contact of the Mediterranean with the central Indian Ocean existed through South West Asia and northern India until sometime in the first half of the Miocene (Fig. 2). Even then, there were clear differences on the species level between the Mediterranean-South West Asiatic

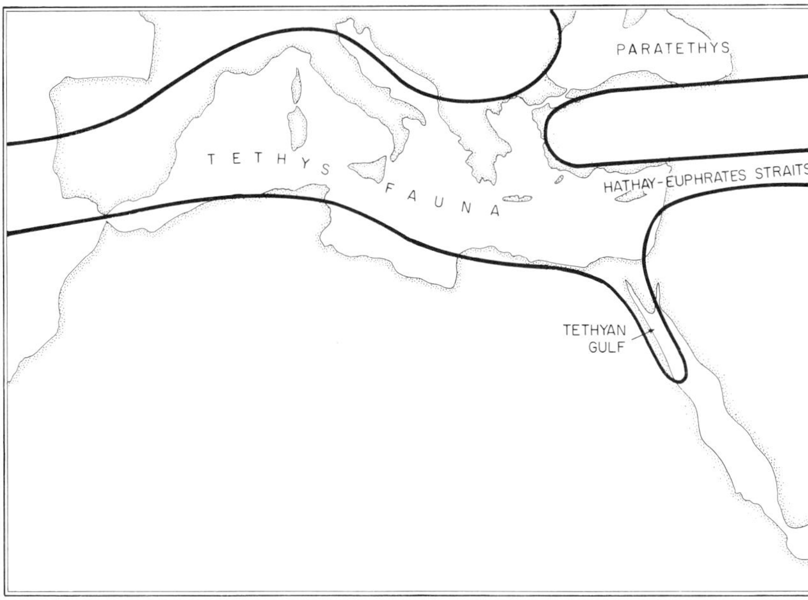

Fig. 2. Approximate map of Tethys Sea in the Mediterranean and Red Sea area (original)

portion and the Malayan portion. Adams (1967, p. 213) who calls the Mediterranean region (only!) the "Tethyan region", states "... the faunas of the Tethyan region and the Indo-Pacific have always been recognizably different in detail ...". This author emphasizes that the faunal movements then were rather in the direction from the Mediterranean to the Indo-Pacific, while the species of this latter region "were quite likely to remain restricted to the one province". The present-day impressive westward expansion of the Indo-Pacific fauna would thus be a relatively new phenomenon.

When and where did the closure of the Mediterranean towards the east occur? To geologists, the Mediterranean is something of a puzzle: "It has the physiographic attributes of a small ocean; its underlying crust is transitional between continent and ocean, and its fauna is typical of that of a sea" (Benson and Sylvester-Bradley, 1971, p. 63). There is no doubt that with the increased application of the theory of plate-tectonics, the whole history of the Mediterranean will be better understood.

The contact to the east was not through the present-day Red Sea graben. Nobody after Issel (1900) thought of an Indo-Pacific connection through the Nile valley. The concept of Hume (1916) and Beadnell (1924) that there was an intermingling of Mediterranean and Indo-Pacific fauna in the Plio-Miocene of the Red Sea, is not supported by the new facts.

Thus, the contact has to be sought in the east or north east. It is generally understood to be a "short way" contact through the Hathay-Euphrates divide of Syria-Mesopotamia. This was a contact already closed by the time of the Lower Miocene, according to Ruggieri (1967). The alternative contact through the

Caspian region was almost concomitantly obstructed according to Benson and Sylvester-Bradley (1971) by the "Paratethys", the inland sea stocked with brackish fauna. [The Paratethys was the first name for the semi-individualized northeastern brackish satellite sea of the Mediterranean, which evolved into the present Ponto-Caspian, and it is not the correct name to describe the still tropical but already isolated Mediterranean of the Miocene-Pliocene, as used by Klausewitz (1968) and Tortonese (1969). The name "Palaeomediterranean" should be used instead for the newly separated, but still tropical, Mediterranean, whether mid-Miocene or Pliocene.]

Adams (1967) criticizes the statement of Ekman (1967) that a considerable disorientation of the climate and the consequent dying out of tropical fauna had already occurred in the Mediterranean during the Miocene. Analysis by Reiss and Gvirtzman (1966a) of the Miocene Saqiye beds of Israel indicates a tropical reef fauna in the late Middle Miocene (Tortonian) and especially the presence of the tropical foraminiferan *Borelis melo curdica* (Reichel) (Reiss and Gvirtzman, 1966b). From the data of these authors and from the chronology given by Reiss (1968), it would seem that the tropical fauna in the eastern Mediterranean disappeared no earlier than seven million years ago with the start of the short Messinian period, considered as the Upper Miocene. Benson (in press, a) however, considers that there was no climatic change during the Upper Miocene. On the other hand, Benson (in press, b) assumes that a general lowering of the oceanic level occurred in the Tortonian (± 8 million years ago) probably isolating in part the Mediterranean from the Atlantic. Ruggieri (1967) was the first to suggest that towards the end of the Miocene, the Palaeomediterranean lost its contact with the Atlantic. The Mediterranean basin was transformed in a series of brackish to hypersaline lagoons and it seems likely to this author that the marine Miocene fauna was totally destroyed (Fig. 3). This state, characterized by *Melanopsis* faunas in the western Mediterranean, would be contemporaneous with the Messinian (Upper Miocene) regression in the eastern Mediterranean (Reiss and Gvirtzman, 1966a). This would have to be considered primarily as a result of the northward movement of the African block against Europe.

The start of the Pliocene coincides, according to Ruggieri (1967), with the opening (or reopening) of the two straits between Atlantic and Mediterranean: the Betic and Rif straits. The Mediterranean was repopulated with marine fauna from the West African coast which was "probably the true asylum for the Indo-Pacific relicts during the salinity crisis of the Upper Miocene" (Ruggieri, 1967, p. 286). If this theory is true, then there is no true Tethys relict in the Mediterranean. Several authors believe that some refugia of marine fauna might have been in existence nonetheless. Benson (in press, b) considers that only very euryhaline species of the "Caspian" type could have survived. The latest data resulting from the Deep Sea Drilling Project (DSDP) in the Mediterranean indicate that by the end of the Miocene there was a long period of repeated "saline crises" and drying up of all of the Mediterranean. Nesteroff (1973b) speaks of "periodic drying ... and transformation in a series of basins". These periods, during which gypsum and evaporites accumulated, alternated with periods during which "these basins became again true seas". This epoch of alternating flooding and drying out is situated, according to Nesteroff, between 7.2 and 5.5 million years, in the

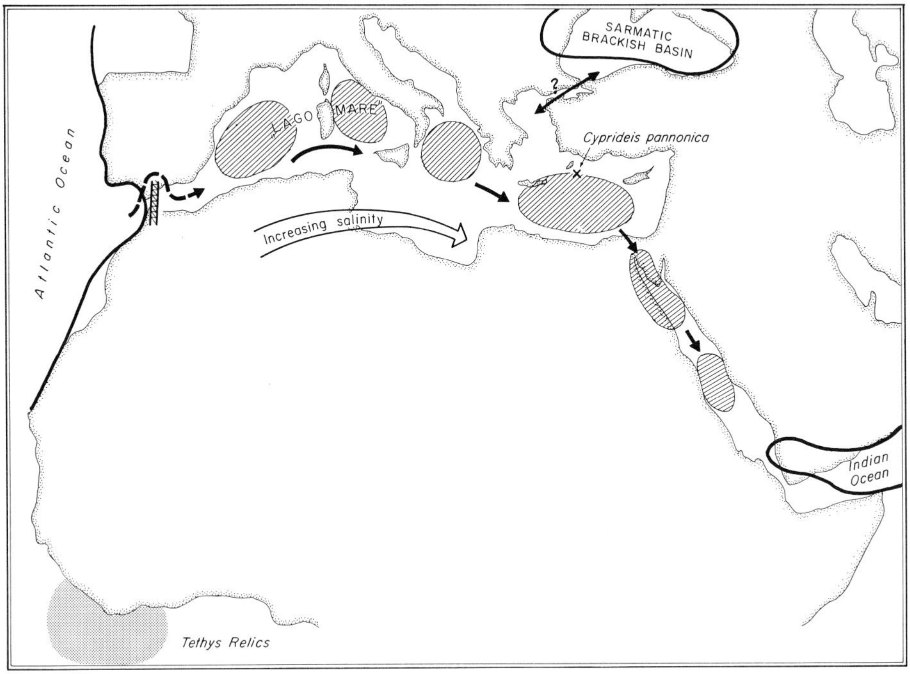

Fig. 3. Sketch map of late Miocene (Messinian) conditions in the Mediterranean and Red Sea area (original)

Messinian period. Benson (1973), and Benson and Ruggieri (1974) add further micropalaeontological data concerning the Messinian salinity crisis and the Pliocene reopening of the Mediterranean to the world ocean. The ostracode *Cyprideis pannonica* is, according to these authors, a characteristic indicator species for the existence of a "lago-mare", a sea-lake or several closed evaporative basins, in the Messinian Mediterranean. Of much interest to our subject is the finding of *C. pannonica* in the core of DSDP Site 120 in the area of Rhodes deep in the Levantine Basin. Lately, more sites with *Cyprideis* were found in the Eastern Mediterranean, and Benson (in press, b) considers that the whole Eastern Basin was under the influence of the brackish Paratethys.

As a result, during the Late Miocene the whole Mediterranean was the site of salt swamp conditions reminiscent of the African "schotts" or the Arabian "sabkhas". Layers of evaporative salt as thick as 2 km were left in the Mediterranean basin which was emptied by evaporation after having contact with the Atlantic (Hsu and Ryan, 1973). According to Benson and Ruggieri (1974), the start of the Pliocene is marked by the reopening of the Atlantic contact and the "extremely rapid, perhaps catastrophic" reestablishment of normal marine conditions. The authors consider this event to be contemporaneous with the *Sphaeroidinella* datum, i.e. five million years ago.

The "Senegalian fauna" of the tropical–subtropical West African coast is therefore the true offspring of the Tethys fauna which repopulated the Mediterranean in early Pliocene. This is also the recent opinion of Klausewitz (1973).

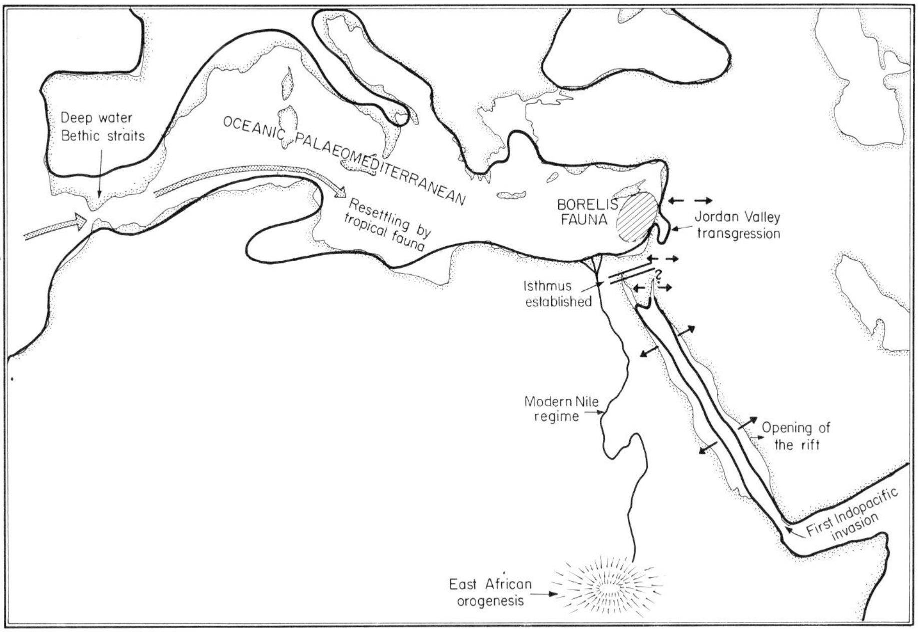

Fig. 4. Sketch map of Early Pliocene deep-sea conditions in the Mediterranean and Red Sea area (original)

This is an important point, since the West African tropical connections of the Mediterranean have been repeatedly, since then, the source of tropical–subtropical elements settling the Mediterranean. Even nowadays the gap between the two provinces is very small. It is generally admitted that the Mauritanian fauna, extending from Gibraltar to Cap Blanc [or the subtropical East Atlantic subregion of Klausewitz (1968)] intergrades smoothly with the tropical Senegalian fauna which is fully established around the Cape Verde Islands. Tortonese (1973a, p. 41) recognizes, following Maurin (1968), a small transitional "Saharian" region (between Cape Bojador and Cap Blanc) "... where the fauna modifies gradually: the relative number of temperate and tropical elements is changed in the favour of the latter".

The early Pliocene, with open deep-sea connections to the Atlantic (Fig. 4), was a period in which the Palaeomediterranean was chiefly inhabited by tropical–subtropical east Atlantic immigrants. According to Blanc (1969, p. 391): "The Pliocene fauna corresponds to that of a warm sea in which late Miocene species exist ... 50% of the Pliocene species have actually disappeared; some, however, still exist in the tropical environment of the west coast of Africa". Genera of tropical type (*Clypeaster*, *Cypreacassis*, etc.), are limited, however, to the Lower Pliocene. The extinction percentage of the molluscs—as compared with the present fauna—increased rapidly from 40% in the Lower Pliocene to 75% at the end of the Pliocene (Ruggieri, 1967). Along the coast of Israel, the Lower Pliocene (Tabianian) was still characterized by some tropical molluscs such as *Strombus coronatus*, *Mitra striatula*, *Cassis laevigata* etc. The tropical elements disappeared in the Middle Pliocene (Plaisancian) (Moshkovitz, 1963, 1968). For the end of the Pliocene, Reiss (1968) gives a radiometric date of nearly two million years (precisely 1.85) and states that the Plio-Pleistocene boundary in the Mediterranean, established by integrated analysis of multiple criteria, preceded the first major glaciation in the Mediterranean area by more than a million years. The end of the Pliocene must have been more or less contemporaneous with the establishment of the sill connection through the recent Gibraltar Straits and the transformation of the Paleomediterranean with its deep-sea connection into the present-day sill-enclosed Mediterranean (Benson and Sylvester-Bradley, 1971). This change set the scene for the Pleistocene environmental fluctuations of this sea.

Much less is known about the Tertiary Red Sea. Nevertheless, events were probably far more simple in this rectilinear graben dominated by the effect of the Syro-African rift movement. In the Eocene, the Gulf of Suez was a southward extension of the Tethyian Mediterranean, similar to the gulf which extended further west into the future Nile valley.

There is disagreement in the evaluation of the age of the last marine Mediterranean influence in the future Red Sea basin and also on whether the Mediterranean and Indo-Pacific faunas ever came to mingle there during the Miocene or Pliocene.

Gohar (1954) assumes that the Paleomediterranean Gulf extended along the present-day Gulf of Suez, and further down approximately to the latitude of Qosseir. After a temporary loss of contact during the Oligocene, the graben was again flooded with Paleomediterranean water in the Miocene, a view accepted by Ekman (1967). Klausewitz (1960) considers that the connection was soon interrupted and that the graben was filled in the Upper Miocene by an "immense inland sea". This lacustrine period is accepted also by Gohar (1954). Brackish water marls and gypsum in the graben are attributed to the Middle Miocene by Hume (1916) and Beadnell (1924).

What happened subsequently is a matter of considerable dissent. The Red Sea of the Pliocene was, according to Gohar (1954), connected to the Mediterranean again but opened for the first time to the Indian Ocean, and consequently acquired a mixed fauna. This conception was taken up by Fox (1926). Klausewitz (1960) extends this coexistence of both faunas to the Lower Pleistocene. Ekman (1967) sets the start of the faunal intermingling at the Middle Pliocene. All these judgments are probably based on the view that the "*Ostrea-Pecten* series" of Beadnell situated by the first author in the Mio-Pliocene and by the second in the Upper Miocene were considered remnants of a mixed fauna. Neumann (1966) writes that sediments along the Gulf of Suez show that the northern end of the Red Sea was open to the Mediterranean only until early Quaternary times. Cox (1929) and especially Moshkovitz (1968) consider that this is probably untrue and that the only marine influence after the Mid or Late Miocene inland lake was an Indo-Pacific one. Moshkovitz also uses the data of Goldberg (1963), from the Island of Tiran at the entrance of the Gulf of Aqaba. He states very categorically that: "In the end of the period of gypsum deposition, the opening of the Red Sea to the Indian Ocean occurred (probably in the Lower Pliocene) and Indo-Pacific fauna invaded. From this time up until now there was no renewed contact between the Red Sea and the Mediterranean. All the proofs for such a contact are more than scanty". Benson (personal communication) considers that all the assumed Pliocene connections between the Mediterranean and the Northern Red Sea are based on misdated Miocene fossils.

With the development of new coring techniques, evaporitic layers similar to those of the Mediterranean were discovered also in the Red Sea. In the Gulf of Suez (El Morgan) this layer is 728 m thick, in the southern Red Sea it reaches nearly 4000 m (after Stoffers and Ross, 1974). The recent DSDP cores established that the Red Sea evaporitic layers are contemporaneous with those of the Mediterranean and are therefore Messinian (Fig. 3). According to Stoffers and Kühn (1974), the Mediterranean and Red Sea had a common history and formed a continuous hydrographic system which went through periods of desiccation caused by the isolation from the Atlantic. Coleman (1974) also states that the Miocene marine invasion came from the Mediterranean. This situation of "sabkha"-like environments in the Red Sea was interrupted in the Late Miocene or very Early Pliocene when the Red Sea opened to the Indian Ocean and normal marine sediments were reported (Stoffers and Ross, 1974). The Pliocene opening of the proto-Bab el Mandab marked the start of the sea-floor spreading which formed the present rift of the Rea Sea. One could assume that the start of this considerable tectonic activity somehow closed the contact to the Mediterranean. Whitmarsh (1974), based on DSDP material, considers that any post-Miocene connections with the Mediterranean were short-lived.

A last point needs to be discussed therefore, and this is the early history of the Isthmus of Suez, which probably appeared in the Upper Miocene. According to Picard (1943) and Swartz and Arden (1960), this was due to a tectonic upheaval. Moshkovitz (1968) makes a very important point when he calls attention to the fact that the Nile sediments, contributed very much to the closure of the Isthmus. Following the Upper Miocene upheaval of the Ethiopian Highlands, the Nile—until then a relatively small river—started to carry huge quantities of silt suspensions which coincided with the interruption of the contact between the Paleomediterranean and the Gulf of Suez branch.

From this one may conclude that both the Mediterranean and the Red Seas underwent hypersaline to brackish periods during the Upper Miocene, which most probably exterminated preexistent fauna. The Paleomediterranean was repopulated in the Pliocene with tropical West African elements, whereas the Red Sea received the first influx of Indo-Pacific fauna.

The "totally dissimilar" [Ekman's (1967) quotation marks] fauna of the two seas before the opening of the Suez Canal in 1869, has its origins in a completely separate evolution from the Miocene onwards. The question as to why the high eustatic Pleistocene sea levels did not result in a "pre-Lessepsian" intermingling of the two faunas will be discussed in the next chapter.

1.2 The Pleistocene of the Eastern Mediterranean

Since at least the start of the Pleistocene [1.8 million years ago, Reiss (1968)], the Mediterranean ceased to be an environmental and faunistic whole. In the words of Moshkovitz (1968) the Eastern Mediterranean and especially the Levant basin were a periphery of the Western Mediterranean.

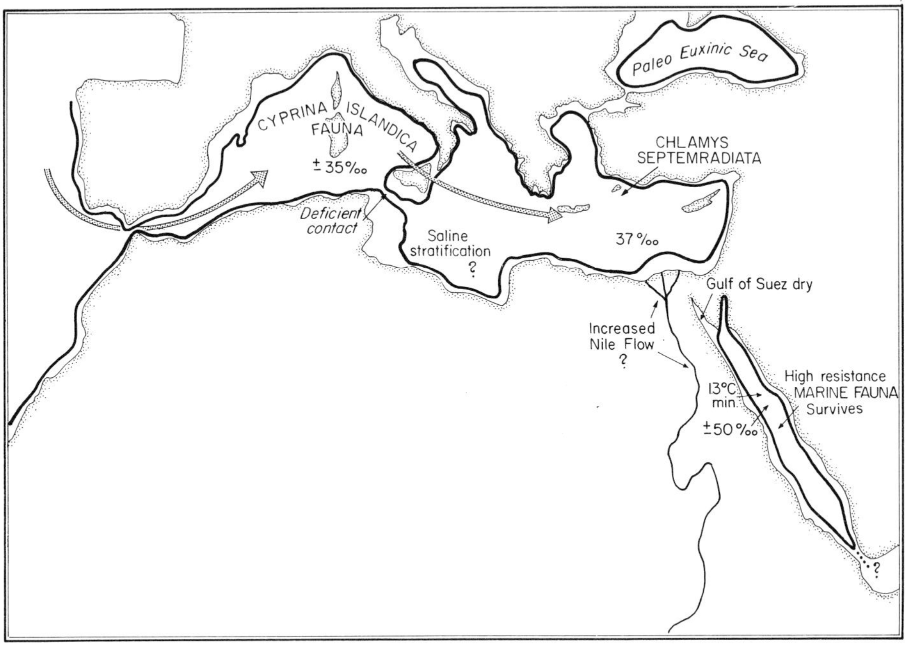

Fig. 5. Sketch map of glacial, low-eustatic sea level conditions in the Mediterranean and Red Sea area (original)

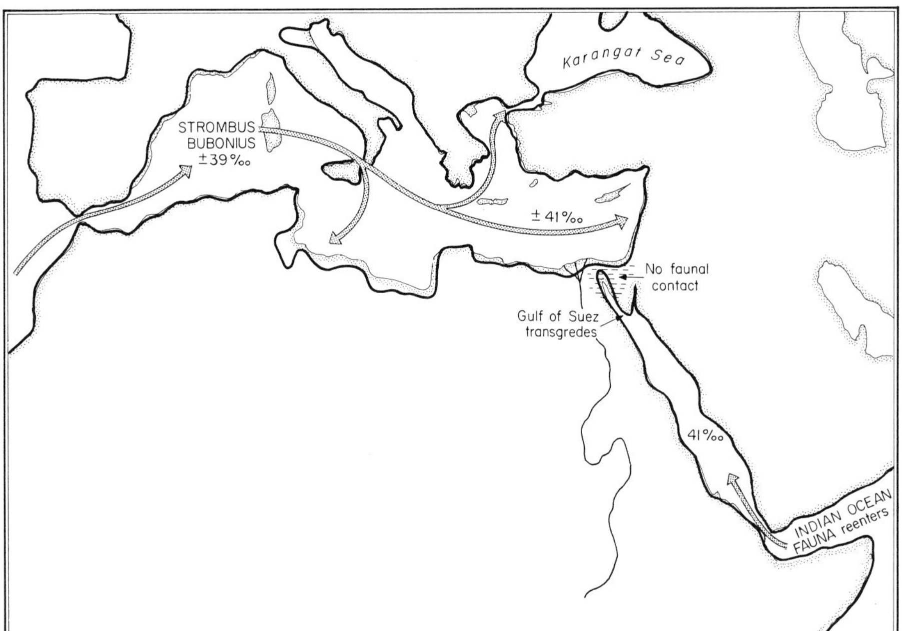

Fig. 6. Sketch map of interglacial, high-eustatic sea level conditions in the Mediterranean and Red Sea area (original)

For the duration of the Pleistocene there is no indication of a positive difference to be found in this region: out of the big diversity of Pleistocene molluscs there is no species peculiar to the eastern basin which cannot be found in the western one. On the contrary, there are many species which never reached the eastern basin. Other species, frequent in the west, appeared only episodically in the east.

The Western Mediterranean—between Gibraltar and the Straits of Sicily—is the classical region in which the universal chronology of the marine Pleistocene has been established. Such terms as Calabrian, Sicilian, Tyrrhenian—indicating different levels reached by the eustatic movement of the sea level are in wide use today and authors like Zeuner (1959) and Fairbridge (1961) see in them a universal pattern.

It is well accepted that during the glacial periods—with much water captured by the increased ice caps—the level of the world ocean was low, perhaps as low as −200 m (Fig. 5). On the contrary, during the interglacials with the melting of the caps, sea levels rose, reaching perhaps as much as +100 m during the earliest of the three interglacials. It was expected that temperature fluctuations would be expressed jointly with the level fluctuations. This, however, is not the case for the Western Mediterranean; only the last interglacial high level (the Tyrrhenian sensu Blanc, 1969) and that of the two last interglacials (Tyrrhenian I and II as in Zeuner, 1959) show an influx of warm water-fauna (Fig. 6).

Zeuner (1959) pointed out that the Pleistocene sea level fluctuations are deviations in a general regressive trend which started with the great late Pliocene withdrawal of the sea. Other authors emphasize that fossil beaches found at high levels inland may be eustatic as well as "tectogenic", i.e. due to tectonic uplifts. This is the reason why a correlation between the fossil beach levels and the glacial stages cannot be seen until the great or second interglacial. The period since can reasonably well be assumed to be tectonically quiet, at least in this part of the world.

Mars (1963) and Ruggieri (1967) emphasized for the first time that an inversion of the currents in the straits of Gibraltar might be responsible for the appearance or disappearance of cold-water or warm-water species coming into the Mediterranean (Fig. 7). During glacial periods, the water balance (runoff/evaporation) was positive in the Mediterranean and therefore there was an outflow of surface water through the straits of Gibraltar and an inflow of deep Atlantic water over the sill. This inflow brought in the "Celtic" cold water elements (Mars and Picard, 1958) which characterize the Calabrian, Sicilian and Würm (last glacial) periods. In the interglacials, evaporation prevailed over runoff and therefore the currents at Gibraltar were reversed: the Atlantic inflow is at the surface and brings in warm-temperature and sub-tropical Atlantic species, whereas the outflow of Mediterranean water is over the sill. The cold-water Calabro-Sicilian species are therefore characteristic of deep water conditions, the subtropical Tyrrhenian species of shallow water conditions. Instead of strictly alternating they could therefore coexist—unless hydrographic conditions were too extreme; indeed they did so.

In the Calabrian and Sicilian species there is a concomitance of "Celtic" molluscs of the *Arctica islandica* association with Pliocene Paleomediterranean,

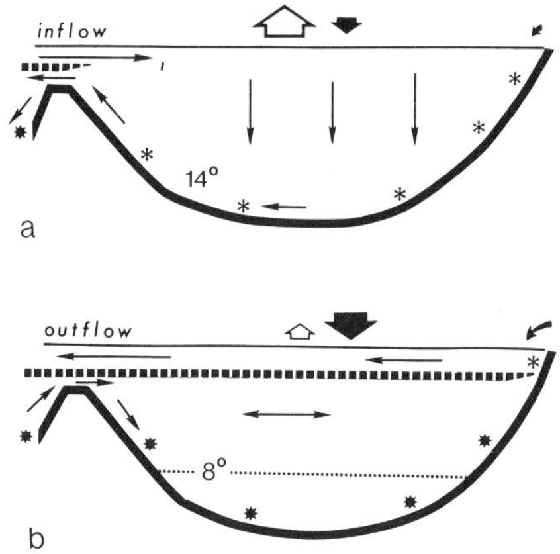

Fig. 7a and b. Scheme of inversion of currents in Pleistocene Mediterranean (from Benson and Sylvester-Bradley, 1971)

presumably warm-water species. Only after the "Pre-Tyrrhenian faunal break" (Zeuner, 1959) did the situation change. During the third (Riss) glaciation, or possibly during the second, the paleomediterraneans were exterminated. Therefore in the following interglacial a warm-water fauna of different origin appeared. This was the typical Tyrrhenian fauna characterized by the snail *Strombus bubonius* and its associates, which are today restricted to the "Senegalian" province, the subtropical shores of Western Africa.

Turning now to the particular case of the Eastern Mediterranean, many questions must be raised.

Zeuner (1959) found a nice sequence of high Pleistocene shorelines in Arabs Gulf, west of Alexandria: ten high beaches from 103 m to less than 5 m. High shorelines were found by Wetzel and Haller (1945), also in the Lebanon, as high as 20 m. However it would be very risky to rely only on eustatic sea level fluctuations in this area.

According to Blanc (1969), the tectonic rise of Rhodes occurred only in Post-Calabrian times—towards the end of the Lower Pleistocene. The downthrust movements in the Jordan–Dead Sea Rift Valley were very active during the earlier stages of the Pleistocene. Reiss et al. (1971) admit a "further deepening of the warm sea" i.e. a tectonic movement as opposed to eustatic movement, as late as the last Glacial/post Glacial boundary, 12,000 years ago.

Furthermore, it is difficult to correlate Eastern Mediterranean high shore levels with the historical phases of the Western Mediterranean, since the "Celtic" cold-water, or better, low salinity association is not represented in the Eastern Mediterranean. There is only one isolated find of *Arctica islandica* from Rhodes (Zaccaria, 1968). *Chlamys septemradiata* is known from the Aegean Sea (Mars and Picard, 1958) (Fig. 5).

The Tyrrhenian warm-water complex is, however, better documented in the Eastern Mediterranean and *Strombus bubonius* was also found at Rosh Haniqra (Israel), at a +5 m level (Moshkovitz, 1968) (Fig. 6).

Alternation of cool and warm conditions in the Eastern basin could be established on grounds of sediment cores and analysis of the fossil planktonic foraminiferan shells (Parker, 1958; Reiss et al., 1971). The dating goes back to the Riss-Würm or last interglacial and a change from warm to cool and again to warm conditions could be proved—especially on grounds of the direction of coiling of the foraminiferan *Globigerina truncatulinoides*. It was easier to synchronize the fluctuations of the planktonic foraminiferan assemblages between Eastern and Western Mediterranean than those of the benthic molluscs. This might be due to the recurrent near-isolation of the Eastern Mediterranean at the Straits of Sicily: such a semi-isolation might have been very significant in hindering the eastward advance of a benthic assemblage, but of no significance for a planktonic population.

Pfannenstiel (1960) considers that at a presumed Riss glacial sea level of -200 m, the Straits of Sicily were only 300 m deep and a few km wide. Following the results of the "Albatross" (in Petersson, 1957) Pfannenstiel reaches the conclusion that at this time "the water exchange between the West and East Mediterranean was practically severed", and that the bottom underwent a stage of deficient oxygenation because of lack of exchange and the saline stratification of the Eastern basin. This stratification was due to the increased amount of runoff during the glacial–pluvial period. In the subsequent Würm glacial, the sea level was at -90 m according to Pfannenstiel—and consequently the isolation of the Eastern basin must have been less extreme.

With all these data in hand we can now return to the initial assumptions of Moshkovitz (1968) which opened this chapter.

The course of Pleistocene events in the Eastern Mediterranean was different from that in the Western Mediterranean. This was a result of late tectonic activity and the intercalary obstacles for water exchange with the west, and eventually also of the changing runoff regime of the major fresh-water supplier, the Nile. The Eastern Mediterranean and especially the Levant basin were an impoverished appendix of the Mediterranean during the whole of the Pleistocene, more isolated from the ocean, with much more environmental instability, fluctuating salinity and oxygenation but always with higher temperatures than in the West. This is the reason why the only phase in which the faunal picture is homogenous over all the Mediterranean is the Tyrrhenian phase. Only then, for the first time since the late Pliocene, can we speak of an even faunal exchange between the two parts of the Mediterranean.

But the Tyrrhenian climax was upset by the Würm glacial. After its first part (Würm I) there was eventually an improvement and a return to Tyrrhenian conditions: the Epi–Monastirian transgression of Zeuner found both in Arabs Gulf (Halig al Arab west of Alexandria) and the Lebanon at 4 m. But the last two glacial stadia (Würm II and III) followed close one after another and the "Senegalian" molluscs characteristic of the Tyrrhenian phase disappeared. Pérès (1967) suggests that some species characteristically found in the southern and eastern parts of the Mediterranean, such as the snails *Fissurella nubecula*, *Cypraea lurida*, *Purpura haemastoma* and *Mitra fusca* and the decapods *Salmoneus jarli*, *Athanas amazone*, *Micropanope rufopunctata*, *Maja goltziana*, and *Pachygrapsus transversus* are relicts of the tropical West-African influx. These species are absent from the Western and Northern Mediterranean but known along the West-African subtropical and tropical coasts. This statement should find support in the fossil findings of the above-mentioned mollusc species. It seems obvious, however, that at least some of the "Senegalian" invaders could in fact survive the Würm glacial.

There is almost no proven case in which a faunal addition from the nearby Red Sea occurred during the whole Pleistocene. If there were such immigrants, they left no traces in the fossil record. The chronically impoverished Eastern Mediterranean received its faunal supply exclusively from the west. The highest elevation on the Isthmus of Suez is only 23 m but one cannot exclude eustatic situations in which the Mediterranean communicated with the Red Sea. High shores at such altitudes are known both from Egypt and from the Lebanon. Fluviatile terraces of the Nile also indicate high Mediterranean levels. It will be necessary therefore to take a closer look at the Pleistocene of the Isthmus of Suez in order to understand why there is no sign of an interchange between the two seas at high eustatic levels.

To conclude this chapter, however, emphasis must be laid on the fact that the "Lessepsian" period of the Mediterranean Quaternary—a period beginning with the opening of the Suez Canal—is the first and only phase in the whole Plio-Pleistocene complex where a massive faunal inflow from the Indo-Pacific occurred. The 45[1] species of molluscs which have already invaded the Eastern Mediterranean can give us a measure of the geological significance of this man-made dispersal event. Future geologists might look at it one day as a reversal of a basic paleozoogeographic pattern perhaps as old as the whole of the Neogene Period, spanning not less than some 20 million years.

1.3 The Pleistocene of the Red Sea

The Red Sea, isolated from the Mediterranean at least since the earlier stages of the Pliocene, had an equally agitated Pleistocene history—along different lines, however.

The movements which led to the opening of the Red Sea Rift were especially active during the Pleistocene and for practical purposes they may be considered as still continuing.

Friedman (1968) considers that along the Sinai coast of the Gulf of Aqaba some minor tectonic movements occurred as late as 5000 years ago. A raised coral reef near Elat was radiocarbon-dated to 4770 ± 140 years, (Friedman, 1968). Hume (1916) described raised Pleistocene coral reefs at an altitude of 200 m—much above any possible high eustatic terrace. The sequence of high and low eustatic levels, which is such a useful tool for understanding the Pleistocene history in the Mediterranean and elsewhere, is therefore of little use in the Red Sea. It can also be assumed that in the Straits of Bab el Mandab, which connect the Red Sea to the Indian Ocean, tectonic activity continued throughout the period and the depth and width of the sill area also changed.

---

[1] Note added in proof: More species are added to the list by Barash, Al., Danin, Z.: Additions to the knowledge of Indo-Pacific Molusca in the Mediterranean. Conchiglie, Milano **13** (5—6), 85—116 (1977).

When, about 30 years ago, the evidence for the important eustatic sea level fluctuations was generally accepted, conclusions for the Red Sea were immediately drawn. Sewell (1948) expressed the view that the sill (about 100 m deep) of the Hanish Islands (north of Bab el Mandab) was dry in glacial times and that the Red Sea became isolated from the Indian Ocean. Under such circumstances the author assumes that two hypersaline lakes were left in the basin and all the marine fauna died out. The Red Sea, according to Sewell, was repopulated with marine biota at the end of the Pleistocene. Klausewitz (1960, 1964) is inclined to consider complete annihilation of the fauna through interruption of the oceanic connections for the first two glacial periods only, but does not exclude the possibility of a late, Pleistocene reestablishment and permanence of the marine biota. Emery et al. (1969) accept the fact that there were in the Pleistocene long periods "during which the Red Sea basin was occupied by a large isolated lake that underwent progressive evaporation". Emery and his co-authors consider that the leaking out of these brines, which were also probably sun-heated, gives the source for the hot brines which are reported from the depths of the Red Sea. Neumann (1966), however, is inclined to place this brine basin in pre-Quaternary times. The latest data of DSDP (see p. 15) give full support to this view.

Gohar (1954) already expressed doubts regarding the hypersaline Pleistocene extermination of the Red Sea fauna, simply on the grounds that the sequence of the coral reefs does not show such an interruption. This author expresses his conviction that there was a "continuity of life in the Red Sea at least from the Pleistocene period until now". This seems to be true albeit in a much modified version.

Emiliani (1961) summarized the then-known data about the amplitude of Pleistocene climatic cycles at low latitudes. Based on O^{18} ratios in foraminiferan shells, he calculated glacial-interglacial temperature fluctuations of 7–8° C in the Caribbean and 3–4° C in the Equatorial Pacific. He also generalized on the work of several authors stating that at low latitudes the glacial climate was arid rather than "pluvial".

A clear picture of the last 80,000 years of the Red Sea—i.e. from the Riss-Würm Interglacial to recent times—was obtained by micropaleontologists studying cores of the hot brine area. Hot brines of up to 56.5° C and a salinity of 257‰ have been found at depths of approximately 2000 m around parallel 21°. Analysis of the cores did little to explain the origin of the hot activity. Nevertheless, the start of the hot extrusions could be set by Deuser and Degens (1969) at 11,000 years BP (Before Present) and by Emery et al. (1969) at 12,000 BP.

On the other hand the micropaleontological material collected in the deep cores of the "Chain" in the fall of 1966 gave a clear picture for the period extending from about 80,000 BP to the present. Of special importance were cores 118 K and 119 K in the hot brine area and core 154 P about 100 miles south. Coccolithophorids, dinoflagellates, radiolarians, foraminiferans and pteropods were analyzed in the cores and the data have been summarized by Deuser and Degens (1969) and by Berggren (1969). Besides the radiocarbon dating, O^{18}/O^{16} ratios for paleotemperatures were analyzed as well as some microchemical data.

The results indicate that between 80–50,000 BP, a period roughly extending between the Riss-Würm or Eemian Interglacial and the early Würm or Pleniglacial, surface water temperature probably varied between 21°–30° C, similar to present-day values in the area. In the two successive periods temperatures decreased until a minimum of 13°–14° C was reached for a short period between 23,000–13,000 BP (Berggren, 1969). About 11–12,000 BP, an abrupt change in climate occurred, which led to "the general type of climate that persisted to the present time" (Deuser and Degens, 1969).

The low temperature periods coincided with periods of lowered sea level or tectonic uplifts in the Southern Red Sea and during these periods the water exchange with the Indian Ocean diminished or even occasionally stopped. The response to that was increased salinity (measured by higher O^{18} content in the water) which, because of the lack of watersheds draining into the Red Sea, could not be diluted by the eventual increase in the amount of rains during the glacials.

From the data of Butzer (1966) it appears that the glacial periods of the high latitudes were not necessarily accompanied by increased precipitations in the subtropical zone. The case of the present-day Persian Gulf, amply supplied with the fresh water of three big rivers at its far end—the Euphrates, Tigris, and Karun—but with open-sea salinity of around 40‰, shows that high evaporation and isolation from the open ocean are of overwhelming importance.

There are no indications as to the amount of salinity increase. This has to be extrapolated from the alternation of stenohaline and euryhaline planktonic organisms. It seems likely that euryhalinity and eurythermy were the important survival factors during the changing oceanographic conditions of the Upper Pleistocene. There are few indications that a true cold-water assemblage ever appeared. Only for the coccolithophorids is there an indication (McIntyre, 1969) that during the height of the Pleniglacial (20–10,000 BP) cool-water species like *Umbilicosphaera mirabilis* and *Cyclolithella annulus* appeared. As for the salinity it reached such high values in late Würm that even the most euryhaline foraminiferid *Globigerinoides ruber* disappeared and the sediments were for a short period devoid of foraminiferida. However, the euryhaline pteropod *Cresseis acicula* survived even under these extreme conditions (Berggren and Boersma, 1969) indicating that conditions remained marine and did not deteriorate to those of a saline brine water (Fig. 8).

Deuser and Degens (1969) emphasize the fact that during the brief disappearance of the foraminiferids from the hot brine area, they did survive in the area of core 154 P, about 100 miles south of the hot brine hole. They assume that the increase in the salinity "was probably stronger the greater the distance from the Gulf of Aden". But since the periods of extreme environmental deviation favored the eurytopic neritic species, the distance from the shoreline also had a marked influence on the biological populations.

It is interesting to remark (see Berggren, 1969) that the uplift of the coral reefs on the Island of Museri, dated at about 17,000 BP by Horowitz (1967), might be corroborated by the core data to show that tectonic movement was also responsible for the extreme isolation of the Red Sea about 20–15,000 years ago.

The present situation was established about 12–11,000 years ago and since then the fauna has not shown show major oscillations (Fig. 9).

One should mention briefly that the relatively high percentage of endemic species of the Red Sea—siblings of ancestral Indian Ocean forms—can be ex-

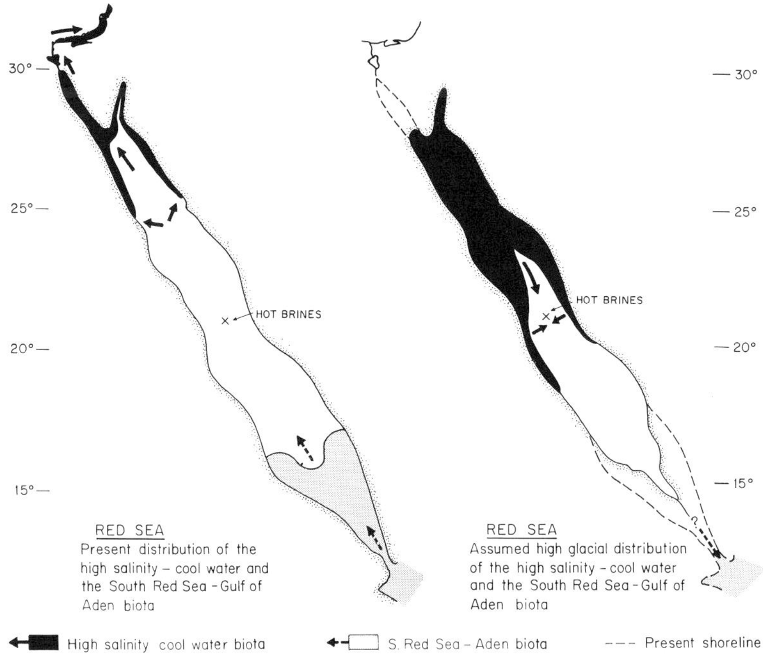

Fig. 8. Glacial and interglacial biotic movements in Pleistocene Red Sea (original)

plained by survival under conditions of near-isolation and the stress of extreme environmental fluctuations (Klausewitz, 1968; Por, 1973b). According to Klausewitz, 10–18% of the fish fauna is endemic and similar rates of endemism are reported for the Decapoda and Crinoidea.

If we consider that the gradient of increasing salinities and decreasing temperatures from Bab el Mandab to Suez which exists today was present and even more marked in extreme glacial conditions, then we have to seek the representatives of the most adaptable fauna and flora in the Gulf of Suez (Por, 1973b). According to Morcos (1970), salinity increases by 4‰ for the 16° of latitude of the main Red Sea and a further 2.5‰ (or even 3‰ mihi) for the 2° of latitude of the Gulf of Suez.

Nothing precise is known about the Pleistocene history of the Gulf of Suez, a shallow (maximum depth 60 m) northern appendix of the Red Sea. Considering, however, the range of eustatic fluctuations, it is presumed that the whole of the Gulf was occasionally dry. At intermediary situations, the shallow sills which subdivide the Gulf into a number of basins might have isolated a number of highly saline basins. These have been hypothetically drawn by Por (1971b) and named "Bitter Lakes" since they must have resembled the situation of the recent Bitter Lake of the Isthmus of Suez as against the Gulf of Suez. But this point will be further developed in the following chapter.

To conclude, it will be useful to quote Berggren (1969, p. 334) who wrote that the Red Sea "has shown itself to be an excellent 'laboratory' for the study of relatively rapid changes of short duration in the marine environment".

Having evolved under such conditions and out of the great diversity of Indo-Pacific biota, it is only natural to assume that the fauna and flora of the Red Sea (especially of its northern part) became very adaptable and thus highly competitive. This is one of the main reasons for its success in the encounter with the Mediterranean biota.

1.4 The Pleistocene and Postpleistocene History of the Isthmus of Suez

One of the most amazing aspects of the subject under study is the fact that an isthmus some 160 km wide with a maximum elevation of 23 m and built up entirely from sediments and sedimentary rock could act as a barrier between the Mediterranean and the Red Sea fauna for at least the whole of the Pleistocene and a considerable part of the Pliocene. At high eustatic sea levels one has to assume that the Isthmus of Suez was flooded and a continuous aquatic connection existed between the Mediterranean and the Gulf of Suez. Conversely, at low eustatic levels, with the whole of the Gulf of Suez dry, at least an additional 250 km was added to the separating dry land. Nothing is known about tectonic movements in this area, which is unfortunate, since obviously a vertical change of even a few meters could be of considerable importance. On the other hand, if the Isthmus can be seen as a product of the depositional activity of the Nile, then the regime of this big river was of primordial importance in the history of the Isthmus. For example, increased depositional or erosional cycles played a major role. The present flood regime of the Nile was established around 12,500 BP (Fairbridge, 1972).

In fact, what is now considered the Isthmus of Suez is part of the old Delta of the Nile. As the main flow of the Nile gradually shifted from the Eastern Delta branches to the Western or Rosetta and Damietta branches, two old branches were gradually left dry by the river: the Wadi Tumilat branch, which flowed due east and emptied in the present Lake Timsah, and the Pelusian branch, which had a northeastern course and emptied in the Gulf of Tineh (Gulf of Pelusium) some 30 km east of Port Said. The Sirbonic Lagoon (Sabkhat el Bardawil), which now occupies about 200 km of the north Sinai coast, was doubtlessly fed directly at times by Nile Delta waters. A third dried-up branch—the Tanitic branch—opened in the area where Port Said stands today.

One body of knowledge relates to the history of the Nile fluviatile regime. Butzer dedicated several writings (1951, 1959, 1966, 1971) to this subject. It appears that the high fluviatile terraces or the cutting in of the river were determined in the upstream region by changes in the amount of precipitation in the East-African mountains and the changing connections with the tributaries, while in the downstream and delta region the main influence was that of the eustatic sea levels. Nile terraces indicate a +35 m level in the Delta which Butzer assigns to the Tyrrhenian, and also a +10–+15 m level which corresponds to the Second Tyrrhenian or Riss-Würm Interglacial. Of utmost importance to our subject is the finding of a +11 m gravel terrace at el Abassa on the Wadi Tumilat Delta branch. One thus has to assume that in the last interglacial the Isthmus was eventually covered by water to a depth of about 10 m. If, therefore, one can assume that during the Mindel-Riss Interglacial the Isthmus was completely submerged (+35 m!), in the Riss-Würm a narrow section of a few kilometers around El Guisr and El Ferdan (Km 64 to 72 of the Canal kilometer count) must have remained dry. Recently, and following other authors, Fairbridge (1972) concluded that the highest Mediterranean interglacial levels might have been only around 20–30 m above the present main sea-level.

The drying-out of the Eastern Delta branches can be followed historically. This is especially the case in the Pelusian branch. Judging from the history of the town of Pelusium (Tel Farama) situated at the opening of this branch to the Mediterranean, the branch must still have been flowing in the earlier Middle Ages or at least in Roman times (Sneh and Weissbrod, 1973). It still appears under the name of Bahr Beni Menaga in the data of the Arab geographer Ibn Sirapiun. The branch of Wadi Tumilat served already in historical times only to carry the overflow of the Pelusian branch into Lake Timsah and the Bitter Lakes basins.

The Tanitic branch probably disappeared earlier than the Pelusian one, but remnants of it still continued to carry Nile waters into Lake Menzaleh. It is interesting to mention that another important Nile branch, the Sebennitic branch, (which flowed exactly in a South-North direction) dried out in the time between the classical times and the Arab era.

A second body of evidence results from the geological prospecting of the Isthmus by Fuchs (1878) in 1864–1865 and the study on the Pleistocene subfossil molluscs collected on the Isthmus by Issel (1869) and Fischer (1870). In the Pleistocene sediments of the Isthmus—prior to the flooding of the Suez Canal—a considerable number of recent species of molluscs was found (Fig. 9). Three groups were recognized by the above authors: Red Sea species, Mediterranean species, and fluviatile, Nile River species.

The Red Sea molluscs found were, among others, *Strombus tricornis*, *Murex anguliferus*, *Fusinus marmoratus*, *Cerithium erythraeonense*, *Mactra olorina*, *Circe pectinata*, and *Ostrea forskalii*. The Mediterranean species found were *Purpura hae mastoma*, *Murex trunculus*, *Cerithium vulgatum*, *Donax venusta*, and *Ostrea edulis*. The Nile species were *Physa contorta*, *Vivipara unicolor*, *Spatha rubens*, and *Etheria semilunata*.

These three fossil faunas are distributed as follows: the Mediterranean species extend over the Isthmus southward to El Firdan (about 64 km south of Port Said) among deposits of sand, sandy clay and gypsum. The Red Sea fauna extends northward from Suez to Kabrit, a distance of about 40 km. The central 60 km exhibit freshwater fauna or an interbedding of freshwater and Red Sea species.

The sill of El Guisr—at an elevation of 23 m—has freshwater deposits and a similar substrate is also found in most of the Lake Timsah area. However, as one advances south, the freshwater fauna is more and more interlaminated with an increasingly rich Red Sea fauna. The Red Sea fauna becomes dominant in the basin of the Great Bitter Lake and exclusive in the Little Bitter Lake.

The most obvious interpretation of the Fuchs-Issel-Fischer findings would be that of a slightly more transgressive Mediterranean and Red Sea and an interposed lagoon area with brackish stratification [fresh at the surface and marine on the bottom as suggested by W. Steinitz (1929)] in contact with the Red Sea (Fig. 10).

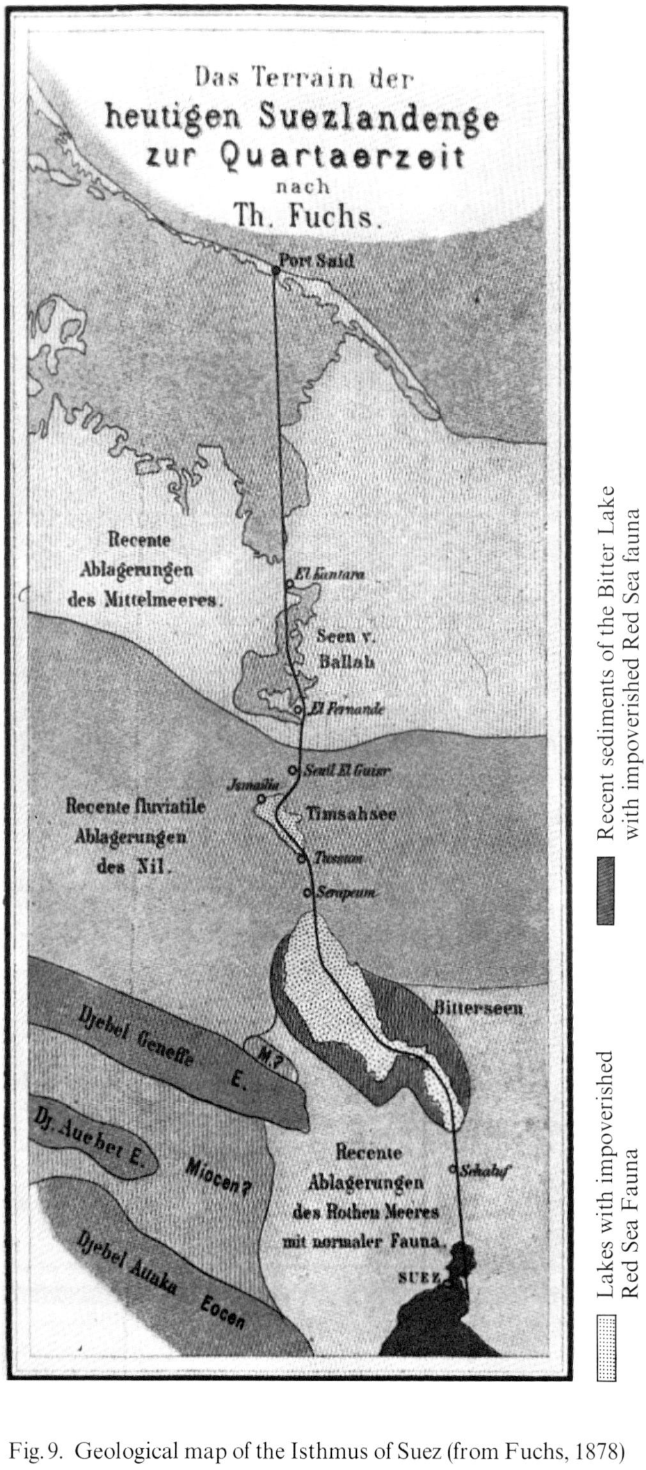

Fig. 9. Geological map of the Isthmus of Suez (from Fuchs, 1878)

The assumption of a Pleistocene strait-like contact between the two seas, leading through a brackish lagoon-like area, was first proposed by Vassel (published in 1889) and accepted by Fuchs (1881) and by Keller (1882). These authors compare the hypothetical situation of the Isthmus of Suez with the Gulf of River Amur, the shallow Straits of Tartaria which separate the Island of Sakhalin from the Siberian mainland. These narrow straits receive the full load of the Amur freshwaters and become therefore a very effective barrier for faunal exchange between the Seas of Okhotsk and of Japan.

In order to assume that the Isthmus area was flooded with Nile water, one has to accept that the eastern Nile branches were very active at those times. The Pelusian branch was carrying water until late Roman times (it is still mentioned by the historians Strabo and Diodorus Siculus, in the 1st Century A. D. and by the Arab geographer Ibn Sirapiun. As for Wadi Tumilat, it carried water—possibly through engineering maintenance—until the end of the 8th Century and even afterwards still brought Nile flood overflow to the Isthmus lakes.

The +8 m deposits of recent type Red Sea molluscs found by Fuchs 5 km east of Kabrit on the Bitter Lake probably indicate an interstadial level during the Last Glacial.

For the post-Glacial climatic optimum (the Atlantic optimum, called also the Flandrian transgression) Butzer (1959) indicates a +4 m level of the Mediterranean. This level existed, according to him, around 6000–5000 BP. It receded to +2 m around 4500–3500 BP, and at this time the Mediterranean shore line was still over 50 km inland from its present stand at the farthest point of Lake Ballah. Considering that the ridge of Shallufa, which separates the Red Sea from the Bitter Lakes, is only +3 m, there is considerable evidence that the Gulf of Suez transgressed in Flandrian-historical times over a stretch of some 70 km of the present Isthmus (Fig. 10).

Krukenberg (1888a) maintains, based on data from Strabo, that at the time of the biblical exodus (13th Century B.C.?) the Red Sea still reached the Serapeum, i.e. 58 km north of its present shore, and that by 500 B.C. it still stood at Shallufa, 12 km north of the town of Suez.

It is also evident from Strabo's writings (1.3) that the recession of the waters from the Isthmus of Suez must have occurred in historical times. Strabo speaks of two possibilities: the increase in the width of the Isthmus which gradually separates the "Egyptian" sea from the Red Sea, or the appearance of a connecting strait owing to the lowering of the Isthmus.

The findings of Lesseps (1876a) who described an alternation of several layers of salt and marl in the Bitter Lake deposits suggest that even after the interruption of the permanent contact with the Gulf of Suez, the basin of this lake received occasional floods from the Red Sea and from the Nile.

Around 400 B.C. the Mediterranean reached, according to Butzer (1959), an extremely low level of −2.50 m. Even if this figure seems too high, one cannot dismiss the fact that ruined towns of the Ptolemaic period (3rd–2nd Centuries B.C.) are some 2 m below the present level of the swamps in the area between the Canal and the Sirbonic lagoon (the Romani area). Contemporary hellenistic scientists of Alexandria, the famous opponents Eratosthenes and Hipparchus, discussed the role of land rising versus lowering in the genesis of the Isthmus of Suez, as well as

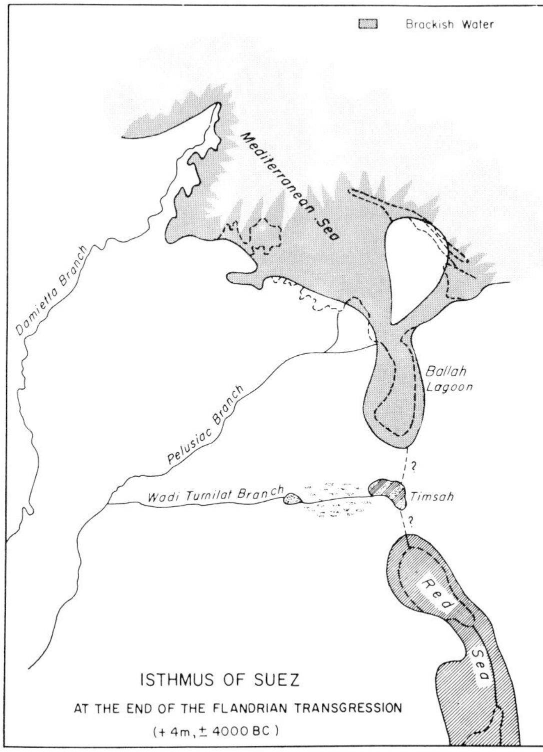

Fig. 10. Isthmus of Suez at the end of Flandrian transgression (from Por, 1971 b)

the reasons for the lack of tides in the Mediterranean versus the relatively high tides of the Gulf of Suez. The two seas must have been already separated at that time.

The present level was reached relatively recently. According to Butzer (1959) this level was attained in the 2nd Century A.D. Sneh et al. (1975) maintain that during the 1st Century A.D. the Mediterranean shore was still up to 12 km inland in the area of Pelusium. Some of the Delta lakes—especially Lake Menzaleh—through which the Suez Canal has been built, were flooded by the returning Mediterranean only in 961 A.D., according to the Arab encyclopedist Makhjumi (Butzer, 1959).

On the "Weimar" map of 1424 the basin of the Bitter Lake does appear, but this might be an echo of Strabo's "Pikre Limne". Seetzen (1813) mentions seven lakes in the basin of the Bitter Lakes—"where the waters of the Nile might reach after heavy floods". Lake Timsah is reported by Schleiden (1858) to be completely filled up at high Nile floods and regularly carrying water in its northern part. Schleiden speaks also of the waters of the Ballah lagoon, an extension of Lake Menzaleh, which reached south to the Guisr ridge.

The history of the water covering the Isthmus of Suez which I have just tried to reconstruct has, however, been complicated by repeated attempts to build

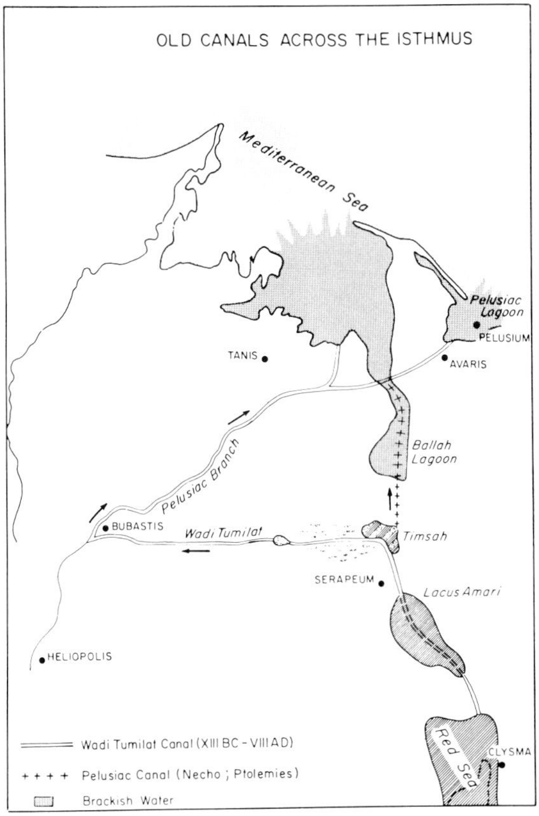

Fig. 11. Old (artificial) canals built across the Isthmus of Suez (from Por, 1971 b)

shipping canals across the Isthmus of Suez. As if trying to keep pace with the gradually widening gap between the two seas and to maintain by artificial means a navigational link which existed since memory, many of the strong rulers of Egypt made themselves known as canal builders (Fig. 11).

According to Strabo the first artificial waterway was built by Sesostris (Ramses II) in the 13th Century B.C. and started near Arsinoe (Suez of today) "in that recess of the Arabian Gulf which penetrates into Egypt. Here are harbors and dwellings and several canals with lakes adjacent to them". The canal went north and then followed Wadi Tumilat to join the Pelusian branch of the Nile near the town of Phacusa. From there, the ships passed into the Mediterranean following the Nile branch to the port of Pelusium.

In the 7th Century B.C., the pharaoh Necho made an attempt to reconstruct the canal which had probably fallen into disuse, but was afraid of the supposed sea-level differences between the two seas and of the possibility of the Red Sea flooding Egypt. Darius the Persian ruler, was more courageous. He celebrated his reconstruction of the canal in the text of a stele on the western shore overlooking the Bitter Lake. It is interesting that Herodotus—although living earlier than

Strabo—does not mention the "Ramses" canal and starts the whole history of the waterway with Necho.

According to Strabo, the voyage through the Isthmus took four days and the canal was broad enough to admit two triremes abreast—i.e. about 30–40 m broad [according to the dimensions of a Greek trireme given by Casson (1959)]. But the triremes had only about 1.5 to 2 m draught, thus the canal could have been very shallow.

About three centuries later, Ptolemeus II Philadelphus rebuilt the canal and added a more direct, northern branch—similar to the modern Suez Canal—which led from Lake Timsah due north through Lake Ballah. By 24 B.C. this canal had also fallen into disuse. There are indications that Trajan rebuilt some sort of a canal, since through the Roman world his name became attached to the canal. Sneh et al. (1975) discovered an ancient canal, unmentioned in historical texts, which ran from Qantara to the town of Pelusium on the Mediterranean shore. These authors consider that the 2–3 m deep and 20 m broad canal pre-existed the Pelusiac delta branch which became active only in the 5th Century B.C. This canal might have been part of a trans-isthmian waterway, although Sneh et al. (1975) consider it more as a sort of strategical water barrier mentioned in the Bible as "Pi-Hirot" or "Shur".

The last canal builder was the conquering Arab general Amru Ibn al Ass, who reopened the Wadi Tumilat connection for use in 640 A.D. The story goes that in the second half of the 8th Century Al Mansur ordered the canal to be filled for political reasons.

The last written record of navigation through the old canal comes from the Irish Abbott Fidelis (7th Century) who reports: "in naves in Nilo flumine, usque ad introitum maris Rubri navigaverunt" (Letronne, 1841).

When Napoleon's engineers made the first measurements on the Isthmus in 1798, they found the remnants of old canal locks 3 m above the present sea level. The engineering efforts of two millenia were thus concurrent with considerable sea-level changes or tectonic uplifts.

1.5 The Problem of the Pre-Lessepsian Migrants Through the Isthmus of Suez

From all that has been said in the previous chapters, it becomes clear that the suspicion of a faunal interchange between the two seas, prior to 1869 which was the year of the opening of the Suez Canal, is justified. The suspicion could have been avoided if we had had complete floral and faunal records for the two seas by 1869. But even now, more than a century later, the records are still incomplete!

True, there is no indication of a pre-Lessepsian faunal interchange in the paleontological mollusc material, but this concerns only one animal group out of many. The micropaleontological material belongs to planktonic organisms which cannot give reliable paleo-distributional information in our case: circumtropical planktonic species could always use the "western way" (W. Steinitz, 1929) i.e. the Straits of Gibraltar, in order to resettle the Mediterranean.

The possibility of a western contact with the tropical West-African areas has also to be taken into account as a further complication.

The following is a list of Mediterranean plants and animals of Indo-Pacific and Red Sea origins which at times and by different authors were suspected to be pre-Lessepsian migrants.

1. *Acanthophora Delilei* (Algae) Aleem (1948), Gohar (1954)
2. *Halophila stipulacea* (Cormophyta) Pérès (1967); Por (1969a)
3. *Ciripathes spiralis* (Antipatharia) W. Steinitz (1929)
4. *Eurythoë complanata* (Polychaeta) Fauvel (1955)
5. *Dasychone cingulata* (Polychaeta) Fauvel (1955)
6. *Pinctada radiata* (= *Pteria occa*) (Mollusca) Por (1972)
7. *Pirenella conica* (Mollusca) W. Steinitz (1929)
8. *Gonodactylus chiragra* (Stomatopoda) Krukenberg (1888a)
9. *Porcellana boscii* (Decapoda) W. Steinitz (1929)
10. *Thalamita admete* (Decapoda) W. Steinitz (1929)
11. *Scottolana longipes* (Copepoda) Por and Marcus (1972)
12. *Metrocarpa nigrum* (Ascidicea) Pérès (1967); Por (1971b)
13. *Hemiramphus picarti* (Pisces) W. Steinitz (1929)
14. *Liza carinata* (Pisces) Ben-Tuvia (personal communication)
15. *Leiognathus klunzingeri* (Pisces) Kosswig (1956)
16. *Aphanius dispar* (Pisces) Kosswig (1967)
17. *Dugong dugon* (Mammalia) Kingdon (1971).

This is a motley list of plants and animals. Some are probably wrong records or misidentifications—as in the case of *Gonodactylus chiragra*—based on an unidentified Mediterranean record by Heller (1863) and Krukenberg (1888), of *Porcellana boscii*, which according to Holthuis and Gottlieb (1958) is based on a doubtful record from the Aegaean Sea by Guérin-Meneville (1832), and the record of *Thalamita admete* made by the same author. *Pinctada radiata*, the little pearl oyster, is most probably circumtropical and *Hemiramphus picarti* too. *Metrocarpa nigrum* could have reached the Tunisian coast where it lives, from the tropical Atlantic. The case of *Halophila stipulacea* has been submitted to serious doubts by Lipkin (1972b). *Acanthophora Delilei* is a case which has not been sufficiently proven and is based only on a short statement by Aleem (1948). *Dasychone cingulata* is, according to Fauvel (1955), very closely related to the Red Sea *Dasychone lucculana* and needs further clarification. *Scottolana longipes* has been considered as "pre-Lessepsian" only on the basis of its present huge populations along the Mediterranean coast of Israel, which for a meiobenthic organism would, in the view I earlier expressed (Por, 1964) exclude a recent invasion. *Leiognathus klunzingeri* seems nevertheless to be a normal Lessepsian migrant. If the presence of *Eurythoë complanata* at Beirut—before the opening of the Suez Canal (Fauvel, 1955)—is a reliable fact, then this species would be a good case of a "pre-Lessepsian" migrant. However, this species is later characterized by Fishelson and Rullier (1969) as circumtropical. Some degree of uncertainty is related to the case of *Dugong dugon*, the Indo-Pacific sea cow, which is considered to have lived in the Eastern Mediterranean at the time of the Phoenician and classical Greek cultures (Kingdon, 1971).

We come now to a very small group among the presumed pre-Lessepsians: *Pirenella conica*, *Liza carinata*, and *Aphanius dispar*, which are characteristically euryhaline and high-salinity resistant species. *Pirenella conica* is a species restricted to the Southern Mediterranean shores and is probably identical with the Red Sea *Pirenella cailliaudi*. The case of *Aphanius dispar* has already been discussed above: this fish is capable of withstanding salinities of around 100%. It is, therefore, much more reasonable to consider it to be a more recent migrant into the Mediterranean rather than a Tethys relict. The grey mullet *Liza carinata* may also be able to resist high salinities but there is no reason not to include it also among the many recent Lessepsian migrants.

Whereas in the case of *Pirenella* and *Aphanius*, their wide distribution in the Mediterranean prior to 1869 makes it necessary to consider them pre-Lessepsians, there is a certain number of species, besides *Liza carinata*, which might have come from the Red Sea and settled restricted lagoon areas along the Mediterranean even before the opening of the Suez Canal. The fauna of the hypersaline Sirbonic Lagoon (Sabkhat el Bardawil, Sinus Sirbonicus of the classical authors) gives some good examples of Red Sea species which can survive salinities such as probably existed in the lakes and pools of the Isthmus of Suez before the building of the Suez Canal. To our knowledge some of these species are still restricted to the Sirbonic Lagoon. The Sirbonic Lagoon will be treated in more detail below.

There is, of course, a slight possibility that some species might have migrated into the Mediterranean during interglacial high eustatic levels and subsequently disappeared. This was suggested by some authors including Kosswig (1956) and Por (1971 b). However, there are no fossil proofs for this and every passing year of intensive research fails to supply any convincing evidence.

At least for the period since the last Interglacial, some 100,000 years ago, there was no stenohaline marine contact through the Isthmus. There were short periods of contact through the freshwaters of the Eastern branches of the Nile Delta. More frequent probably were the cases in which isolated hypersaline pools, left behind by a retreating sea-tongue or a spring flood or else an abundant Nile flood, could carry a certain fauna to survive till the next flood. In such short jumps some very euryhaline species could have spread over the chain of Isthmus lakes: the Bitter Lake, Lake Timsah, Lake Ballah, Lake Menzaleh, and the Sirbonic Lagoon. These basins have probably carried waters of some degree of hypersalinity all the time (see Seetzen's and Schleiden's information above). Lake Menzaleh was probably reduced to a few small salt pools until invaded by the sea in the Middle Ages, when it turned into a brackish basin. The Sirbonic Lagoon shows considerable changes of salinity if the bar-openings to the Mediterranean are open or silted. Both lagoons are indicative of the pre-Lessepsian conditions of the Isthmus water-bodies.

Even when the classical shipping canals were functioning, shallow and narrow as they were (see above), they could not change the prevailing hydrographic conditions of the Isthmus lakes: hypersalinity in normal years and occasional flooding by Nile waters. The shipping canals, with a possible short exception during the Ptolemy period, always led through the Tumilat–Pelusium way, i.e. through the freshwaters of the Nile.

Under such conditions only a few estuarine species would be able to survive and advance. Some of them—like the barnacle *Balanus amphitrite*, or species of the isopod genus *Sphaeroma*—are so widespread that it is hard to say which way they invaded the Isthmus: from the Red Sea or from the Mediteranean. It is interesting in this respect to quote Keller (1882) who states that: "*Sphaeroma serrata* ... vielleicht *nebst anderen* (our italics) sich schon vor Eröffnung des heutigen Suez Kanals in den salzigen Tümpeln auf dem Isthmus vorfanden". According to Glynn (1972), there are actually three species of euryhaline isopods of this category in the Canal: *Sphaeroma serratum*, *S. walkeri*, and *Cymodoce truncata*.

In some species the direction of migration is clear, as in the case of *Pirenella conica*. There is also a clear-cut case of a euryhaline bivalve *Cerastoderma glaucum* (*=Cardium edule*) which came from the Mediterranean, and, before the cutting of the Canal, succeeded in spreading as far as the northernmost end of the Gulf of Suez (see W. Steinitz, 1929).

In two previous papers (Por, 1971a, b) I proposed the term "Isthmus Fauna" for this particular euryhaline fauna, the migrational movements of which might have preceded the Lessepsian canal.

Riemann and Rachor (1972, p. 185) have recently added some nematode species to the list of the Isthmus fauna. They write: „Diese Arten können Reste einer alten Bittersee Fauna repräsentieren, die bereits vor der Öffnung des Kanals bestand". Por and Marcus (1972) also added several species to the presumed list of the Isthmus fauna.

The list of the species tentatively considered to belong to the Isthmus fauna is as follows:

Nematoda:	*Theristus flavensis* „Gruppe" (Sensu Riemann and Rachor)
	Theristus oxycerca
	Oncholaimus oxyuris
	Syringolaimus striatocaudatus?
	Spilophorella paradoxa?
	Ptycholaimellus ponticus?
Polychaeta:	*Augeneriella lagunari*
Copepoda:	*Canuella perplexa*
	Canuellina insignis
	Robertsonia salsa
	Nitocra spinnipes orientalis
	Nitocra affinis
	Heterolaophonte quinquespinosa (nec *Paralaophonte*, Por, 1973a)
	Neocyclops salinarum
	Pseudodiaptomus salinus
Ostracoda:	*Cyprideis torosa*
	Aglaiocypris sp.
Cirripedia:	*Balanus amphitrite communis*
Isopoda:	*Sphaeroma serratum*
	Sphaeroma walkeri
	Cymodoce truncata

Mollusca: *Cerastoderma glaucum* (= *Cardium edule*)
Pirenella conica (= *P. cailliaudi*)
Mactra olorina?
Brachidontes variabilis?
Pisces: *Aphanius dispar*
Liza carinata?
Liza aurata?

Another addition to this list may eventually be the serranide fish *Crenidens crenidens*, known from saline lagoons on both sides of the Isthmus. A comparison of the parasitofauna in the Mediterranean and Red Sea populations of this fish (Paperna, 1972d) has shown it to be completely different. The two populations were thus probably isolated for a period longer than one century and therefore *C. crenidens* might also be considered as a pre-Lessepsian migrant from the Red Sea. It was already reported in 1882 from Lake Menzaleh by Keller.

It seems thus fairly convincing that the pre-Lessepsian movements across the Isthmus were restricted to euryhaline lagoon species which, even if they emerged on the other side, remained more or less restricted to the neighboring shores and lagoons.

There is, of course, also the possibility that Egyptian or Greek craft of classical times, crossing from one sea into the other, may have carried some fouling organisms or other adventives. The cases of *Halophila*, *Pinctada*, *Metrocarpa* (see p. 31) might perhaps be explained in this manner. Lipkin (1972a, 1975) upholds a different view: *Halophila* could easily be transported by ships through the modern Suez Canal, but he excludes for some reason the possibility of a pre-Lessepsian event.

Last but not least, the sea cow *Dugong dugon* was a cultic animal; as for instance, of the Philistine goddess Derketo. An artificial colonization of this animal in the Mediterranean cannot be entirely ruled out.

1.6 The Sirbonic Lagoon (Sabkhat el Bardawil)

Much of the northern coast of Sinai is occupied by a large lagoon known to the classical geographers as "Sirbonis Limne" or "Sinus Sirbonius", mentioned by Schleiden (1858) as "Sirbonissee" and named by the Arabs "Sabkhat el Bardawil". Most of the authors use this last name or the name "Bardawil Lagoon". There is no reason, however, to discard the old name of Sirbonis.

The Sirbonis occupies an area of 650 km^2, 30 km east of Port Said to 20 km west of the town of El Arish. It is on the average 10 km wide, but at one point (Mitzfaq or Nahal Yam) the width is 22 km. The lagoon is separated from the sea by a narrow, low sand bar. At only one point, Mons Cassius (Ras Burun or El Kals), does the bar rise to a height of 5 m. Depth in the Sirbonis nowhere exceeds 1.5 m, and there are many areas where the water is less than 1 m deep. There is only one natural and more or less permanent opening to the sea, at the far-eastern end of the lagoon—the Boghaz Zarniq. A second such opening appears on the maps, somewhat more to the west. Since 1927 (Ben-Tuvia and Herman, 1972), two artificial openings have been established in order to decrease the salinity in the lagoon and allow the annual immigration of commercial fish. The two artificial openings respectively designated as Boghaz 1 and Boghaz 2 are marked on the map in Figure 12.

It seems that before 1927 the whole basin was a "Sabkha" (meaning in Arabic a "salt swamp"). During the winter storms, the gales flooded the swamp (as they do today) and towards the summer the whole area or the majority of it evaporated and turned into a "playa". Before the building of the Suez Canal, which cuts through Lake Menzaleh between two earthen banks, the waters of this lake probably communicated at high Nile floods with the Sirbonis. It can be presumed that the basin became a real lagoon several times during the centuries, carrying water all the year round. It is of interest to give at some length the references of Strabo (1st Century of Christian Era) to the Sirbonis. In his words "when the sea retired, these regions (northern Egypt, *author's note*) remained bare except that the Lake Sirbonis remained; then the lake also broke through to the sea and thus became a bog" (1, 50). In Volume XVI, 760, Strabo gives the dimensions of the lake: 200 stadia in length and 60 stadia maximum breadth, i.e. some 34 km by 11 km, one-third of the present length and half of the present maximum width. The lake communicated with the sea through an opening called Ecregma ("the outbreak") somewhat east of Mons Cassius. But in Strabo's time "the Ecregma has become filled up with earth". At the time of Plinius, about half a century after Strabo, the "Sirbonis Lacus" was already a medium-sized marsh lake ["nunc est palus modica" (5, 68)]. Both Strabo and Diodorus of Sicily reported occasional gales invading the Sirbonis from the sea.

On several old maps such as the 15th Century Weimar map, the 16th Century map of Tileman Stella and the 18th Century map of Tirion and Loveringh in Amsterdam, there appears a definite gulf, called even on the earlier map "Sirbonis Lacus".

Schleiden (1858) used the example of the Sirbonic lake to predict that a future canal through the isthmus would similarly be clogged by sand. In short, conditions in the Sirbonis were not different from the present ones, at least for the last two millenia: temporary contact with the sea and successive periods of isolation with occasional storm gales flooding the marshes.

The Sirbonis is the best actualistic example of an old isthmic hypersaline basin which existed prior to the building of the Suez Canal. For this reason, a chapter in this book is dedicated to the Sirbonis.

Nothing was known in the scientific literature concerning the hydrography and the biota of the Sirbonis before 1967. In 1967 the lagoon was in contact with the sea through three openings. The main opening was closed by sediment during 1969–1970 and only in 1971 was the opening again established and deepened, and former conditions returned. Therefore, on a short time-scale, one could follow to some extent the presumably normal and cyclic changes in the Sirbonis.

Salinity data are based on Oren (unpublished) Ben-Yami (unpublished), and Roth (1972, 1973). The main sea water influx occurs through Boghaz 1 in the west and the other openings are chiefly outlets. Lowest salinities are near the openings and they increase southward to Mitzfaq or towards the dead waters in the gulf or the isolated pools. The annual range of fluctuation is 10–20‰ with a peak in September–October and a low in March-April (Por, 1971a) (Fig. 12a, b). Absolute values fluctuate geographically and seasonally between 40–73‰ (Roth, 1973). When the western opening was closed in the summer of 1970 the salinity in most of the lagoon was 80–95‰. In some remote areas it exceeded even 100‰. With the reopening of Boghaz 1 in August 1971 the salinity became "normal" again.

Summer temperatures may exceed 30° C but, owing to the shallowness, O_2 content never decreases below 80% saturation (Roth, 1973). Nitrate and silica contents seem to be higher than in the open sea (in a yearly average), whereas no difference is found in phosphate values (Roth, 1973).

The flora and fauna of the Sirbonis are qualitatively poor, showing a drastic impoverishment from the openings inwards. Knowledge is still at a preliminary stage and only fishes and benthic copepods were studied in detail (Por, 1972; Ben-Tuvia and Herman, 1973). There is, however, a general knowledge of the main benthic organisms (Por, 1971 b, 1973 a; Ehrlich, 1975 a); and of the dominant plankton (Kimor and Berdugo, 1969). Some further organisms were identified: the sabellid polychaete *Augeneriella lagunari* by Gitay (1970), some of the plants by Lipkin (unpublished), molluscs by Barash and Danin (1972) and the chironomid *Cricotopus mediterraneus* by Margalit (unpublished).

In the plankton, Kimor and Berdugo (1969) recognize several species of Diatomacea among them *Synedra hennedyana*, *S. gailloni*, and *S. undulata*. The species *Campylostylus striatus* is found in high-salinity areas of over 70‰; it has also been reported from salt springs of the Dead Sea area. There are three species of *Ceratium*—all of them euryhaline neritic species also known from Suez Canal waters. There are three species of Tintinnoidea: *Tintinnopsis radix*, *T. beroidea*, and *Favella campanula*. The first two are of Red Sea origin and Kimor and Berdugo (1969) write concerning the third: "... it may subsist though not thrive in the hypersaline lagoon". The planktonic copepods so far identified are the euryhaline species *Euterpina acutifrons*, *Acartia clausi*, and *A. latisetosa*. The two latter species are known from the Suez Canal. Most interesting is the presence of the cladoceran *Bosmina coregoni maritima*—unknown in the Eastern Mediterranean area except the Sirbonis and the Suez Canal (Kimor, 1972). *Artemia salina* lives only in some hypersaline pools isolated from the main body of the lagoon (Por, 1972).

The Diatomacea of the benthos are very diversified, and according to Ehrlich (1975 a) they fall into three categories: (1) the stenohaline marine species are concentrated around the openings and do not live beyond salinities of about 40‰; (2) euryhaline marine species are distributed to a salinity limit of 60‰; and (3) the holeuryhaline diatoms, of continental origins, are found all over the lagoon and become dominant to exclusive at salinities above 60‰. In dead assemblages collected in 1970, the year of maximum salinity due to the closure of the openings, the euryhalines dominated the lagoon (Ehrlich, 1975 a). This author lists 147 species and varieties of diatoms; however, only some 15 species represent the dominant euryhaline assemblage.

This picture is very similar to that found in the benthic Copepoda (Por, 1973 a), although this study was carried out in 1970 when salinity was high and some of the salinity limits observed represented possibly suboptimal circumstances. From among the 15 species found, seven species advanced only as far as salinity values of 45‰. A second group consisting of four species survive to values of 65–70‰. Two species of marine origin are found up to 85‰, namely the harpacticoids *Robertsonia salsa* and *Heterolaophonte quinquespinosa*. At salinities

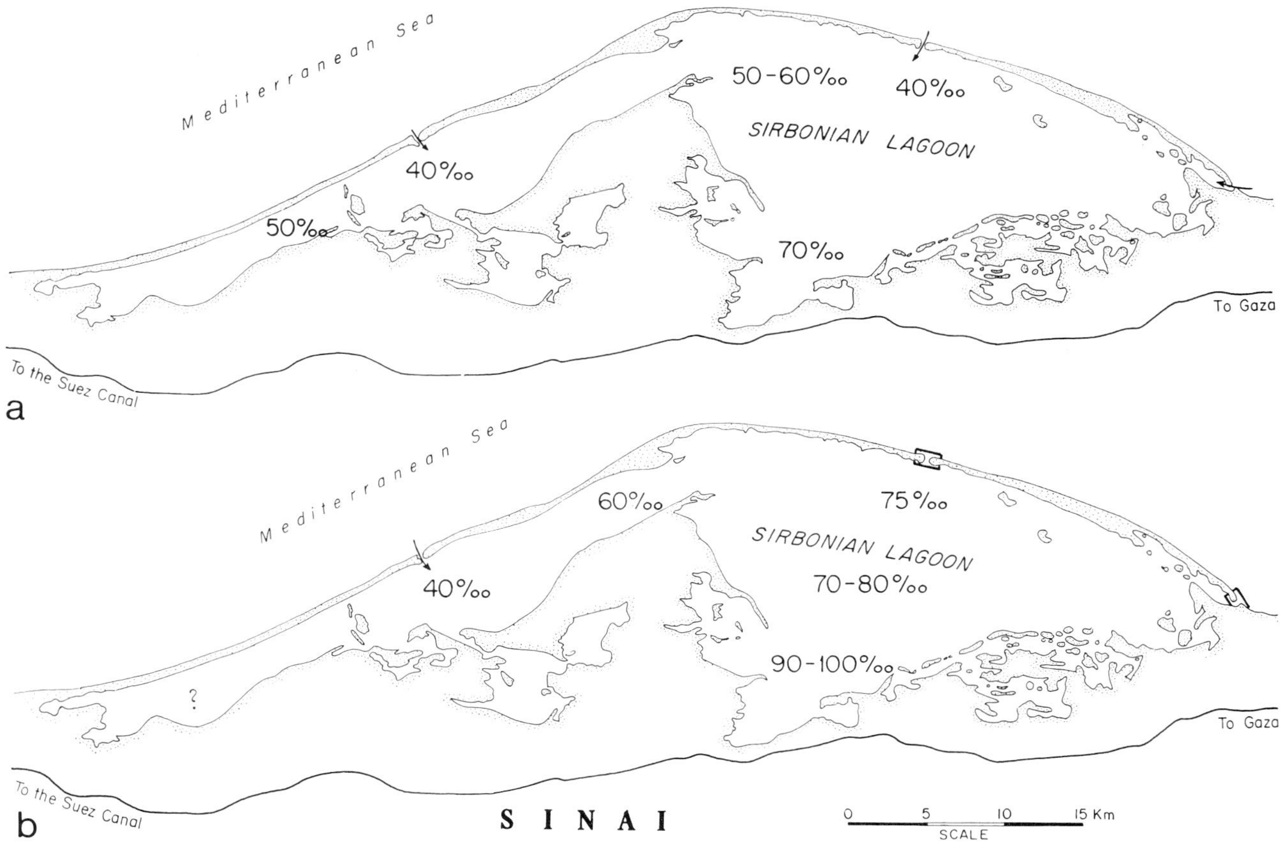

Fig. 12a and b. Schematic distribution of salinities in the Sirbonic Lagoon (Sabkhat el Bardawil) (a) in 1967, when both openings were functioning; (b) in 1969, when only western opening remained active (from Por, 1971a)

over 90‰, a marine cyclopoid *Neocyclops salinarum* meets two euryhaline harpacticoids of continental origins, *Nitocra lacustris* and *Cletocamptus confluens*. It is of considerable interest that 12 of the 15 species are also known from the Suez Canal and its surroundings. Six species are of Red Sea origin and one or two additional species, although possibly circumtropical, also came by way of the Canal (Por, 1973a).

No other group of the zoobenthos has been analyzed in detail. There are two euryhaline ostracods (*Cyprideis torosa* and *Aglaiella* sp.), many nematodes, a watermite, and one or two species of mysids. The larva of the chironomide *Cricotopus mediterraneus* reaches predominance at high salinities (around 60–70‰) and is an important food basis for some fishes (Ben-Tuvia and Herman, 1972).

The macrobenthos is excessively poor. The only macrophyte is the flowering plant *Ruppia ovalis*. On the plants are rich epiphytic growths of *Cladophora* sp. (Lipkin, unpublished). This last species developed especially during the high salinity years of the closure of the openings (Ben Tuvia and Herman, 1972). Only five species of mollusc are reported, namely *Cerithium scabridum*, *Pirenella conica* (or *P. cailliaudi*), *Cerastoderma glaucum*, *Mactra olorina*, and *Brachidontes variabilis*. The first four species live on level bottoms, *B. variabilis* on wooden piers and poles. It is interesting to note that this little mussel can be found even at a salinity of 70‰.

An unidentified hydrozoan and the sabellid polychaete *Augeneriella lagunari* live on the *Ruppia* stems. *Balanus amphitrite* accompanies *Brachidontes* on the wooden structures. A *Sphaeroma* sp. seems to be the only type of isopod present.

The migratory element, reproducing in the open sea and entering the lagoon for a certain period, is represented among decapods and fishes. The shrimp *Metapeneus stebbingi* is a widespread species, but according to Tendler (1972) in the high salinity year of 1970 it was relatively rare at 60‰. The swimming crab *Charybdis* sp. was also reported.

A total of 41 species of fish have been reported from the lagoon by Ben-Tuvia and Herman (1972). This number increased to 65 (Ben-Tuvia, 1975b). Only *Aphanius dispar* reproduces in the lagoon and survives at salinities over 70‰. The other species probably only feed in the lagoon. The most important commercial fish, *Sparus aurata*, was introduced from the Lagoon of Venice by Italian fishermen (Ben-Tuvia, personal communication). The estuarine fishes predominate: *Dicentrarchus labrax* and *D. punctatus*, six species of grey mullet, among them the Red Sea *Liza carinata*, the atherinids *Atherina mochon* (Mediterranean) and *Pranesus pinguis* (Red Sea), the Mediterranean goby *Pomatoschistus marmoratus*, the Red Sea Sparidae *Crenidens crenidens*, etc. It is interesting to note that the fishes of Red Sea origin constitute 25% of the species in the Sirbonis, whereas in the open Mediterranean they constitute only 10% (Ben-Tuvia, 1975b).

In conclusion, the relevance of the study of the Sirbonic Lagoon (or Bardawil Lagoon) for the Lessepsian migration, can be summarized as follows:

1. The flora and fauna of the Sirbonis may serve as a model for an Isthmian, pre-Lessepsian saline waterbody, especially in its periods of isolation from the sea.

2. The wide range of high salinities in the lagoon, once renewed contact to the sea is established, give excellent opportunities for a model of the first stages of the invasion of the Suez Canal when salinity was still high.

2. The Suez Canal

2.1 The Building of the Canal, Its Physical Features

When the first spades struck the Isthmus in 1859, many preliminary measurements had already been made and the international discussion about the feasibility and the future of the Canal was already centuries old.

After the closure of the old canal in the 8th Century, several plans were aired to reconstruct the old waterway. There were plans made in Venice around 1500, and the German philosopher Leibniz, in a memorandum called "Consilium Aegyptiacum" (1671) proposed the building of the Canal to Louis XIV. At the same time there was a permanent and lingering fear that if a canal were built the Red Sea might flood the lowlands of the Nile Delta. This assumption which survived over the centuries is also found in Strabo's account of the old canal project. James Capper in 1783 wrote: "If the canal was to be supplied with water from the Red Sea an extraordinary spring tide or storm might have ... overflowed lower Egypt so as to render it a mere salt water lake". Despite all this, many enlightened minds in Europe among them Goethe and the young general Napoleon continued to play with the idea of rebuilding the canal. During his Egyptian campaign (1789) Napoleon rode with a small escort to Suez to visit the remnants of the old canal. Even in his last years on St. Helena he wrote enthusiastically about the Canal project. He charged the engineer Le Père with the task of investigating the problem of the differences in the level of the two seas. Le Père reached the conclusion that at high tide the Red Sea is about 10 m higher than the Mediterranean. Although the English surveyor Captain F. R. Chesney corrected the measurements in 1830, it took an international team of three engineers, headed by the Austrian Nigrelli, to establish definitively that the maximum difference at very high tides in the Gulf of Suez is only about 1.30–1.50 m. This difference between the tide of the Red Sea and the practically tideless Mediterranean is just enough to create strong tidal streams in the southern part of the Suez Canal and to contribute to a northward flow in the northern part of the Canal. Actually, the sea level at Suez may sometimes reach more than 2 m over that of the Mediterranean.

Schleiden (1858) mentions for the first time the fears that the proposed canal would soon be obstructed by blown-in sand. Ferdinand de Lesseps (1873) tried to refute this fear along with the prognosis of an evaporative drying-out of the Canal. He took these risks, along with many others, and made his achievment also an example of personal daring (Fig. 13). The sanding-in of the Canal was still considered as a real danger when the Canal was closed and not maintained

Fig. 13. Meeting of the two seas in the Bitter Lake, last pickaxe blow on August 15th, 1869 (woodcutting by Edward Riou, from Pudney, 1968)

Fig. 14. Satellite picture of Nile Delta with Suez Canal in background (Gemini IV; courtesy NASA)

between 1967–1974. Although the data are not complete and finally proven, it seems that the changes of depth which occurred in the Canal during these years do not indicate a silting up. It took much arguing and political maneuvering for de Lesseps to obtain his final charter: the British opposing the Canal and using the technical authority of W. Stephenson, the French using all the technico-romantic enthusiasm of the Saint Simonians (the "possibilistes" and the "canal-istes").

In the following pages a few technical data about the Canal will be given, but no complete topographic and physical description of the Canal (Figs. 14 and 15, Map 1).

A total of 75 million m^3 were excavated by 1869. The total length of the Canal is now 162.5 km. To that must be added the western pier of the harbor of Port Said, which has been built 6.5 km out into the Mediterranean, while the eastern jetty of Port Said is only 2.5 km long (Fig. 16). A jetty approximately 2 km long exists at Port Ibrahim at the Gulf of Suez (southern) end of the Canal.

Only about 70 km of the Canal are dug out in dry land, while the rest of the waterway crosses a series of lakes and swamps (from the north: Lake Menzaleh, Lake Ballah, Lake Timsah and the Bitter Lakes).

The 45 km through Lake Menzaleh are deepened in the shallow lake, with the excavated earth forming two dams which contain the Canal (Fig. 17). By this, the eastern end of Menzaleh has been cut off from the main lake and turned into a

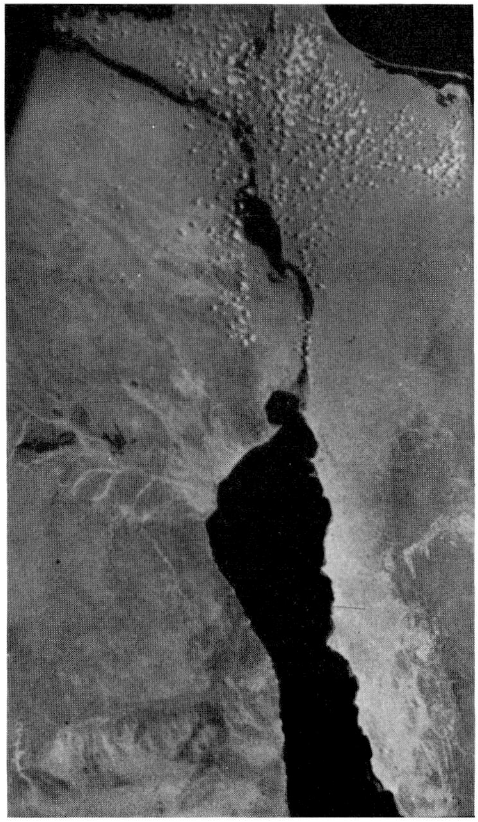

Fig. 15. Satellite picture of Suez Canal (Gemini VI; courtesy NASA)

salt swamp. Lake Menzaleh is a brackish oligohaline waterbody, 1 to 2 m deep, periodically supplied with fresh water from the Nile Delta. A few kilometers south of Port Said there is a narrow channel with a lock which connects Lake Menzaleh and the Canal. The Canal works started in the Menzaleh section. In the very first phases waters of this lake served to fill the first "service channel".

The following 30 km are artificially dug either through the salt and gypsum swamp of Ballah or through a sandy ridge which at El Guisr (km 72) reaches a height of 23 m.

A stretch of 4 km leads through Lake Timsah (Fig. 18). The undredged maximum depth of this depression is 6 m. Before the construction of the Canal the basin was marshy with Nile waters reaching it from time to time at very high floods through the valley of Wadi Tumilat. The lake has been dredged to a depth of 13 m to allow the passage of ships. There are many lagoons and gulfs on both sides of Lake Timsah, some of them fairly deep (3 m). Since the lake receives an inflow of fresh water from various sources it is reported to have a brackish stratification with the heavy salty water overlain by a diluted layer. The marginal lagoons offer, therefore, a wealth of different salinity conditions. The main source

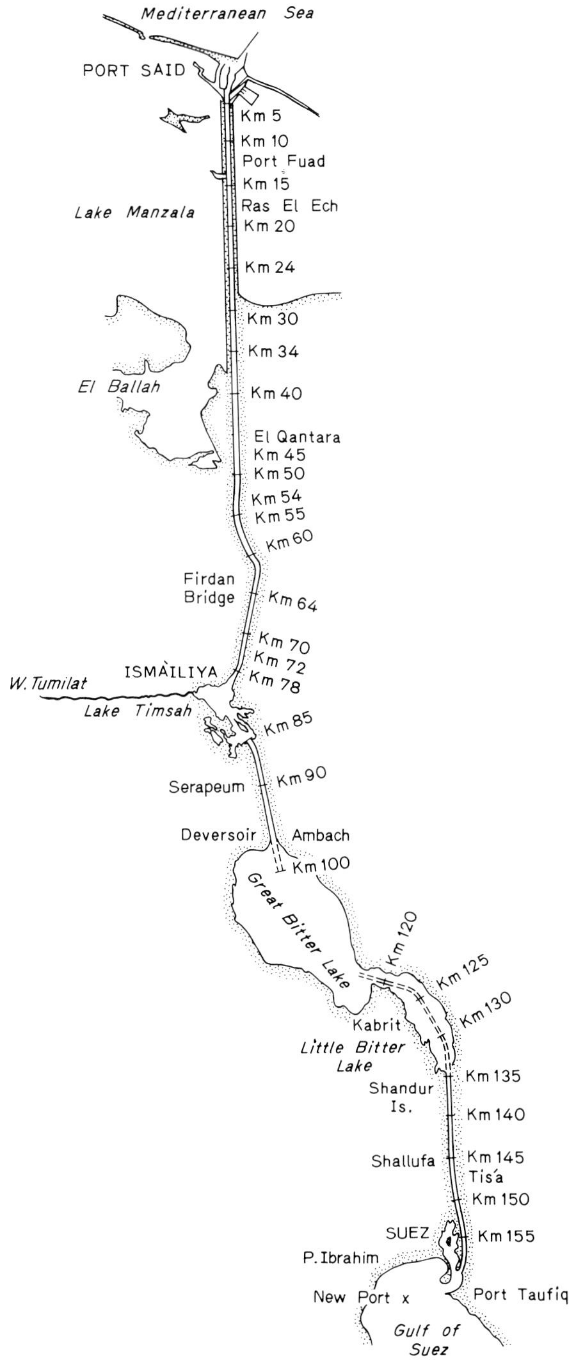

Map 1. Suez Canal

Fig. 16. Aerial photograph of northern section of the Suez Canal, with Port Said in background and Lake Manzala on left (courtesy I.D.F., Jerusalem)

of fresh water is an outlet of the fresh water canal from the Nile which provides drinking and irrigation water to the region. After 1891 the fresh-water canal was continued north and from then on only overflows reached the lake. However, the developing agriculture supplied much runoff. The Mediterranean waters reached Lake Timsah for the first time in 1863, but the filling up of the depression continued intermittently and was finished only in 1867. More will be said about Lake Timsah in the discussion of the biota of the Suez Canal.

Between Lake Timsah and the Bitter Lake there is a stretch of 16 km. The shores of the Canal are strengthened for their whole length with a stone covering. At Km 92 the ridge of Serapeum reaches a maximum height of 10 m, and at some distance along the African shore there are several isolated salt pools (the pools of Nefiche).

The Bitter Lake is the central and most important waterbody of the Suez Canal (Fig. 19). According to Thorson's (1971) calculation, it contains 85% of the waters of the Canal system. The total length of the Lake is 36 km, but at the promontory of Kabrit there is a narrowing to a strait about 2 km wide, which separates the basin into a northern part (the Great Bitter Lake with a maximum width of 13 km) and a much narrower southern part (the Little Bitter Lake).

The basin of the Bitter Lakes, the "Pikre Limne" of antiquity was a "sabkha"-like salt swamp into which very high tides of the Gulf of Suez flooded (Linant, 1860). Water was present especially along the shores where also vegetation (chiefly *Tamarix* thickets) evolved ("La forêt d'Ambach"). According to Aillaud (1868) this "forest", about 7 km long and 2 km wide, was inhabited by a rich fauna of water birds. The center of the depression was occupied by a salt mass 13.20 m (de Lesseps, 1871, 1876a) or even 18.20 m thick (Gruvel, 1936) and covering a

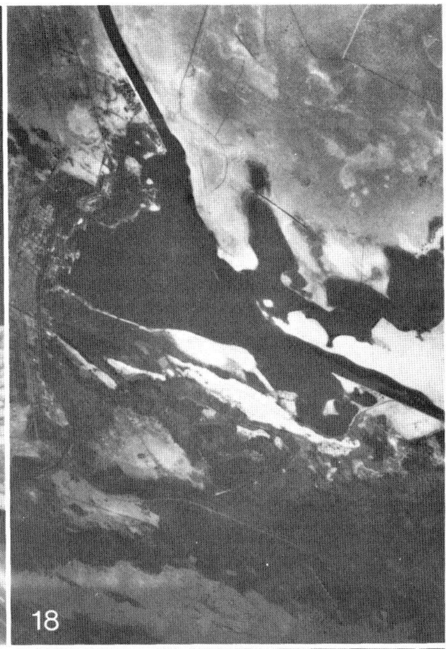

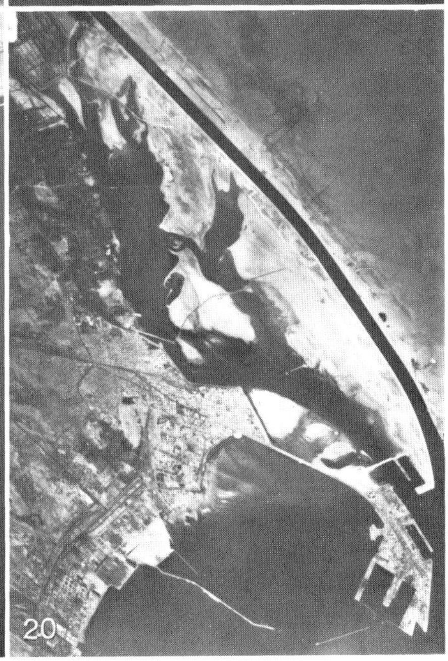

Fig. 17. Aerial photograph of northern end of Suez Canal (courtesy Y. Etam)

Fig. 18. Aerial photograph of Lake Timsah, with Isma'iliya on left and lagoons on right (courtesy Y. Etam)

Fig. 19. Aerial photograph of northern part of Great Bitter Lake, with lagoon of Deversoir (el Ambach); (courtesy Y. Etam)

Fig. 20. Aerial photograph of southern outlet of Suez Canal, with lagoon of Suez (courtesy Y. Etam)

surface of over 100 km^2. The depression in the northern part was over 12 m deep and, therefore, little deepening had to be done in the northern basin. The shores are sandy and, on the eastern shore, with several big salt swamps (Fig. 20). The western shore is steeper and a few rock outcrops are formed; at water level, beach rock is formed. The filling of the Bitter Lakes from the Mediterranean started in the spring of 1869. On August 15th 1869 the waters of the two seas mixed for the first time in the Bitter Lakes (Fig. 13).

The last 27 km to Suez show another important feature—a rocky obstacle, the 2-m-high sill of Shallufa—in the past separating the Bitter Lakes from the Red Sea and causing one of the last-minute troubles for the opening of the Suez Canal in October 1869.

At km 159 a channel leaves westward and turns north to the old city of Suez. In fact the last 3–4 km of the Canal are built in an extension of the Gulf which became dry only in the first decades of the last century and was flooded again by the Canal works (Fig. 20).

The Bay of Suez being very shallow, continuous dredging and deepening of the bay are carried out for about 5 km offshore to the New Port Rock.

At the beginning, the navigational depth of the Canal was 8 m, the surface width 59–98 m, while the minimum bottom width was 22 m. The Canal has been widened and deepened several times. These operations started in 1876 after the first years of successful use. In 1958 after the first stage of the Nasser Project, the depth of the Canal was 13 m, and the Canal width 125 m; the permissible ship draught was 11.5 m. The further stages of the Nasser Project would have been finished in 1972, had they not been interrupted by the 1967 war, and would have brought the depth to 14.7 m and the canal width to a maximum of 250 m. This would have raised the maximum tonnage of the ships permitted to cross the Canal from 40,000 to 60,000 tons.

The completion of the Nasser Project would have brought the total excavated amount of cubic meters to 365 million m^3 (175 millions before the start of the project and 190 millions during the project itself).

Some 20% of the world tonnage passed yearly through the Canal, of which 70% were oil tankers. Big refineries were built west of Port Ibrahim at the Gulf of Suez outlet; already in 1936 Gruvel complained about the deleterious effects of the oil pollution.

The Suez Canal was closed several times during armed conflicts in the area. During the fighting of 1915 navigation was interrupted for only a few days. In the Second World War, the interruption was for 76 days. This increased to several months after the 1956 crisis. After June 1967 the traffic was shut down, to be started again only in 1975. From December 1973 to May 1974 a compact dam, built from one shore to the other slightly north of the outlet of the Canal into the Great Bitter Lake, interfered with the free flow of water through the Canal. By April 1975 the Canal was cleared of all obstacles and in June 1975 it was reopened to navigation.

From information presently available it does not seem that the long period of interruption in the dredgings along the Canal basically changed any of its biologically significant topographical features. Lack of dredging, decrease in turbidity and pollution probably had some favorable effects on the living communities of the Canal; however, comparative data are not available.

The clearing of the Suez Canal from the wrecks left there in 1967 was finished by the spring of 1975. In reviewing the salvage operations carried out in the Canal, Barracca and Thomas (1975) expressed their surprise at the fact that "... the Canal has accumulated little silt in almost eight years of disuse". This becomes understandable in part, since these authors mention tidal currents in the southern sector of the Canal which sometimes reached 4 knots and made it "impossible for divers to work in open water".

2.2 History of Research

At the great World Exhibition of 1867 in Paris, the "Compagnie Universelle du Canal maritime de Suez" exhibited, among other things a rich collection of molluscs from the Isthmus of Suez.

This symptomatic fact indicates how intimately the engineering works of the Canal construction were connected with scientific research. Ferdinand de Lesseps himself published lengthy papers on the geology and hydrography of the Bitter Lakes (de Lesseps, 1871, 1876). Voisin Bey, the first chief engineer of the Canal, published six volumes under the title "Le Canal de Suez" 1902–1904, with much hydrological and chemical information. Other scientific publications were written by engineers, geologists or ships' captains associated with the Compagnie, such as the geologist Fuchs, Cpt. Vassel, or the ichthyologist Tillier, and the malacologist Bavay (see Table 1).

The great scientific importance of the eventual faunal interchange between the Mediterranean and the Red Sea was realized as soon as the Canal project started. Vaillant (1865) collected molluscs in the Gulf of Suez in 1864 in order to have a basis for comparison, after the changes he foresaw: not only species migration but also adaptation of the invaders to the new environment and their hybridization with the autochthonous species. Issel in his "Malacologia del Mar Rosso" (1869) compared the molluscs on both sides of the Isthmus in order to establish what species were common to the two seas before the new man-made contact. The project of Fischer (1865) on the molluscs of the "two shores of the Isthmus of Suez" had the same purpose. The geologist Fuchs, who spent two years (1864–1865) with the Compagnie and then returned in 1876, gave a picture of the subfossil thantocoenoses of the molluscs on the Isthmus.

In the first 30 years of the Canal only few general works appeared; however, owing to the good previous work done on the molluscs, these began to be used as a test-case for animal migration through the Canal.

Two more general inquiries into the whole of the migrational processes were made by Keller who visited the Canal in 1882 (1882, 1883, 1888) and by Krukenberg (1888a) who spent the winter of 1886–1887 there.

There are several reports on the salinities and also other chemical parameters during the first three decades, supplied by the Compagnie, as well as by the throughgoing oceanographic research ship "Pola" [October–November 1895

(Natterer, 1898)]. Morcos (1972) gives a most useful summary of early chemical investigations in the Canal.

The new century started with the fundamental ichthyological work of Tillier (1902) which also contained much environmental information. This was followed by the equally important publication by Tillier and Bavay (1905) on the shelled molluscs of the Suez Canal.

In the years that followed, the appearance of Red Sea immigrants in the Mediterranean was recorded for the first time: Jordan and Hubbs (1917) reported the finding of the Red Sea fish *Pranesus pinguis* at Port Said and Fox (1924) that of the swimming crab *Portunus pelagicus* as far as Haifa.

In 1924 an expedition to the Suez Canal was organized by the Royal Society, known afterwards as "The Cambridge Expedition to the Suez Canal, 1924". The group led by H. M. Fox and comprising R. Gurney, V. C. Robinson, and D. N. Twist, spent the months of October, November, and December 1924 along the Canal.

This was to be the only organized zoological expedition to the Suez Canal; the results which appeared in the Transactions of the Zoological Society between 1926–1929 were to be the only exhaustive body of knowledge on the biota of the Canal. Altogether 37 papers on the diverse animal groups resulted, accompanied by Fox's comments and conclusions. The only other long-term enterprise on the Canal was that of Gruvel and Moazzo (1932–1934) which, besides a general biological and fisheries description of the Canal by Gruvel (1936), produced only the ichthyological papers by Chabanaud (1932–1934), the malacological paper by Moazzo (1939) and the data of Monod (1933, 1937, 1938) on the Crustacea. Since then only short visits to the Suez Canal were carried out by Tortonese in 1944–45 and Beets in 1950.

Two expeditions collected and studied considerable material from the vicinity of the Suez Canal. One of them, the "Mission Robert Ph. Dolfuss en Egypte" between December 1927 and March 1929, collected in the Egyptian Red Sea and had its results published in a number of papers in 1933, 1938, and 1959. The activity of Steuer resulted in a series of taxonomic papers entitled "The Fishery Grounds near Alexandria" between 1935 and 1939.

The appearance of Red Sea migrants along the Levantine coast was studied by W. Steinitz (1927–1933); Monod (1930), Liebman (1935), Haas (1937), and Pallary (1938). After World War II, a series of papers was published by different authors in the "Bulletin of the Sea Fisheries Research Station in Haifa".

The algae of the Canal have scarcely been studied until the recent papers of Lipkin (1972a, b). There are only two small notes: Lyle (1930) based on the Cambridge Expedition material and Lami (1932) on the Gruvel collections. The study of the algal migrants to the Mediterranean was started by Rayss (1941–1963).

The hydrography of the Canal did not serve as a subject of special study from the end of the last century until Wüst's paper in 1934. Since then, Krauss (1958) and especially Morcos (1960–1967) give more continuous and comprehensive information about the hydrography and especially the salinity of the Canal. Muromtsev (1960, 1962) gave some hydrographical data collected by a throughgoing Soviet research ship. Miller and Munns have recently (1974) summed up the oceanographic information yielded by the two throughgoing American vessels, "Atlantis II" and "Chain".

The picture given by this short presentation of the history of the Suez Canal research is rather discouraging (see also Table 1). Except for two periods, the years immediately before and after the Canal opening (i.e. 1864–1888) and the period between 1924 and 1934, there is no comprehensive information to characterize the evolution of the Suez Canal biota. In recent years only the studies of Morcos on the salinity give more continuous information. "The Hebrew University—Smithsonian Institution Joint Program" trips to the Canal (1967–1973) yielded relatively restricted collections. These have, however, been studied fairly completely. A special issue of the Israel Journal of Zoology [Vol. **21** (3–4), 1972] gives a summary of most of this research effort.

In January 1975 Professor Brattström collected some of the fouling which grew on the ships trapped in the Bitter Lake since 1967; this collection has yet to be studied.

A scientific appreciation of the migration process through the Suez Canal has to content itself with this rather incomplete data. In the more than one hundred years since the opening of the Suez Canal, one could have expected a more organized and continuous effort by the scientific community.

Table 1. Chronological listing of scientific research in the Suez Canal

Collector and Author	Year of collection	Year of publication	Topics
Vaillant, L.	1864	1865	Mollusca
Issel, A.		1869	Mollusca
Fuchs, Th.	1864–1876	1878, 1881	Geology, Mollusca
Fischer, P.		1865, 1870, 1871	Mollusca
Müller, K.		1872	Chemical data
Lesseps, F. de[a]		1873, 1876	Geology, chemistry
MacAndrew, R.	1869	1870	Mollusca
Durand-Claye, L.	1872	1874	Chemical data
Schmidt, C.	1875	1878	Chemical data
Keller, C.	1882	1882, 1883, 1888	General biology
Krukenberg, C. F. W.	1887	1888	General biology
Vassel, E.[a]		1889, 1890	General data, geology
"Pola" Expedition Luksch, J., Natterer, K., Sturany, R.	1895	1898, 1899, 1903	Chemical data, Mollusca
Tillier, L.[a], Bavay, A.[a]		1898, 1905, 1906	Mollusca
Tillier, J. B.[a]		1902	Pisces
Voisin Bey[a]		1902–1906	General data
Herdman, Thompson, I.C., Scott, A.	1902	1903	Planktonic Copepoda
"Ammiraglio Magnaghi" Vercelli, F.	1923–1924	1927	Hydrography
Monod, Th.		1925	General
Cambridge Expedition to the Suez Canal: Fox, H. M., et. al.	1924	1924–1931	General; zoological results: reports by specialists on groups of animals

[a] Persons employed by the Suez Canal Company.

Table 1 (continued)

Collector and Author	Year of collection	Year of publication	Topics
Wüst, G.	1924–1926	1934, 1935	Hydrography
MacDonald, R.	1928	1933	Plankton
Gruvel, A., Moazzo, G., Chabanaud, P., Lami. R., Monod, Th.	1932–1934	1932–1939	Plants, animals (Mollusca, Pisces)
Ghazzawi, F. M.	1936	1936, 1938	Plankton
Tortonese, E.	1944–1945	1947, 1948, 1952	General, Pisces, Echinodermata
Beets, C., Holthuis, L. B.	1950	1953	Bitter Lakes: geology, Benthos, Decapoda
Morcos, S. A., Riley, J. P.	1953–1955 1964–1966	1960, 1966, 1967, 1968, 1972	Chemistry
Muromtsev, A. M.	1959, 1960	1960, 1962	Hydrography
Miller, A. R., Munns, R. G.	1963, 1965	1974	Hydrography
Dowidar, N. H.	—	1973, 1974	Plankton
The Hebrew University — Smithsonian Institution Joint Program Steinitz, H., Por, F. D., Aron, W., et al.	1967–1972	1972	Animal and plant groups; migration through the Suez Canal
Safriel, U., et al.	1971–1973	—	Migrating species of Polychaeta, Mollusca
Hecht, A., et al.	1973	—	Hydrography Great Bitter Lake: Benthos (Collections at the Hebrew University of Jerusalem)
Norwegian Council for Research and Humanities Brattström, H., Taasen, J. P.	1975	—	Great Bitter Lake: fouling from ships stuck since 1967 (Collections at the Hebrew University of Jerusalem)

2.3 The Hydrography of the Suez Canal

Judged by the standards necessary to understand such a complicated environment, the hydrography of the Suez Canal is practically unknown. Only lately (Miller and Munns, 1974) has the first attempt been made to build a hydrographical model of the Canal: however, at this date the Canal was already basically different than during its first century of existence.

Salinity is the best-known environmental factor and there are some scattered data about currents and temperatures. Such basic factors as water levels, evaporation, transparency, nutrient contents, and primary production are mostly unknown. Some of these data may still be contained in the log books of the Suez Canal Company but these data have never been fully utilized. Here and there, authors give credit to information obtained from the Canal authorities but it is hoped that more data on the past periods in the evolution of the Canal system may still be available and awaiting publication. As it stands now, the picture of the distribution and dynamics of the hydrographical parameters in the Canal, over the seasons and during its 106 years of existence, is fragmentary.

Therefore, in this field of easily measurable facts, hypothetical assumptions still prevail. A few of these, in more or less historical order are:

1. The hypothesis that the Canal waters were originally less saline and that it took several years until they reached their high salinities.
2. The hypothesis of the silting up of the Canal, unless permanently dredged.
3. The belief that the high salinity of the Bitter Lake is due only to the dissolution of the bottom layer of salt and that this lake will eventually reach normal sea water salinities.
4. The belief that the high salinities measured in the northern Gulf of Suez result from an outflow of Bitter Lake water.
5. The much repeated assumption that the minimum temperatures decrease from the Gulf of Suez, northward through the Canal.
6. The assumption that there are throughgoing currents through the whole of the Canal system.
7. The role ascribed to the Nile floods in generating the seasonal North-South current in the northern Canal section.
8. The belief that there are seasons in the Canal which are especially propitious to the migration of animals through it.

The Currents. There are few measurements of the currents in the Canal. Fox (1926) stated that up until that time the Canal Company had never made a systematic study of the currents, because of the regularity of the tidal currents in the southern Canal and the dependence on the permanent current in the northern section.

Much of what is given in literature about "currents" is based on salinity differences measured at different periods of the year, while changes in the horizontal and vertical value distribution were interpreted as indicative of water mass movements (Wüst, 1935; Morcos, 1960). If temperature is not taken into account, this assumption is very much weakened. The first author to give temperature/salinity relation (T/S) diagrams and density calculations, is El Sabh (1969).

The old data about displacement of water masses in the Canal were based on the assumption that the only source for the highly saline water body was in the depth layer of the Bitter Lake. The possibility of an increase due to evaporation in density in the shallow lagoons surrounding the Bitter Lake and Lake Timsah has not been taken into account. However, it is now evident (see below) that with maximum salinities in the Canal system being around 45‰, as at present, evaporative processes clearly prevail over dissolution processes—if they still occur at all.

Three basic vectors influence the currents of the Suez Canal: (1) the water level differences between the Gulf of Suez and the Mediterranean at Port Said, with the addition of the evaporative level decrease in the lakes of the Canal; (2) the direction and the strength of the tidal waves; and (3) the direction and strength of the winds.

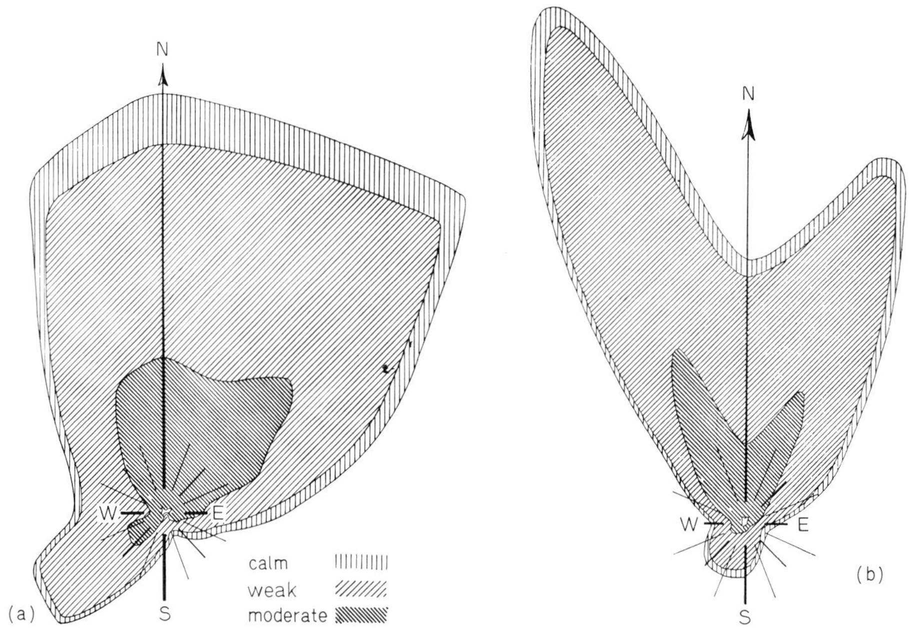

Fig. 21a and b. Wind directions in Suez Canal area (a) at Port Said; (b) at Port Taufiq (from Gruvel, 1936)

The winds, to start with the last of the above factors, blow almost exclusively from North-North-East (Gruvel, 1936) (Fig. 21a and b). Wüst (1934) indicates that some 90% of the winds are northerly. Only in the winter are there some days of southerly winds as a result of the distant influence of the winter monsoon in the Southern Red Sea (Morcos, 1970). By and large, it seems that the influence of the winds is only of secondary importance: they influence only the immediate surface or become important when the level differences tend to become zero.

On the average, the level of the Red Sea at Port Suez is higher than that of the Mediterranean for the major part of the year. Tillier (1902) sets this difference at 30 cm. Wüst (1935) gives for the period October–June a difference of 29–40 cm. Morcos (1960) mentions 36.3 cm. Rouch (1940) and following him, Lisitzin (1965) calculated an average level difference of 24 cm, and later (Lisitzin, 1974) of 20 cm.

This level difference is negligible in the months of August and September. According to Tillier (1902), the Mediterranean stands 40 cm higher than the Red Sea in these months. Lower values are probably more correct: 17.6 cm (Morcos, 1960) and 12.3 cm (Morcos and Gerges, 1974). Wüst (1935) confines himself to saying that in this period the level of the two inlets is equal (see also El Sabh, 1969). The reason for the seasonal equalization of the levels has to be sought in the decrease of the water levels in the Northern Red Sea owing to the increased summer evaporation. Some authors have also attributed it to the massive addition of Nile waters to the Mediterranean in the neighborhood of Port Said, when with open dams 140–180 million m^3 of water gushed out of the nearby Damietta branch of the Delta. However, observations after the cessation of the Nile flow do not support this assumption.

Spring tides at Suez are sometimes as high as 2.1 m (Wüst, 1934) or 1.8 m (Tillier, 1902). Fox (1926) gives a main spring tidal range of 1.50 m. The "Red Sea and Gulf of Aden Pilot" (1955) gives sea level rises of up to 2.7 m during southern winds, which causes strong tidal currents to enter the southern part of the Canal.

Because the Bitter Lake acts as a reservoir, the incoming current is stronger than the outgoing one. Values for the incoming current are indicated as 1.38–2.00 m/sec (Tillier, 1902) or maximum 1.40 m/sec and average 0.82 m/sec (Fox, 1926). Miller and Munns (1974), on information supplied by Morcos, give for July–August average current velocities of 98.19 cm/sec at Shallufa.

The outgoing current is weaker. However, according to Morcos (1960) the outgoing current becomes stronger during the summer than the incoming one, perhaps in connection with the low summer levels in the Gulf of Suez.

Since tides at Suez are semidiurnal, the tidal currents in the southern Canal change directions four times a day. According to the "Red Sea and Gulf of Aden Pilot", the onset of the northward tidal current is 2 h after the high tide at Suez and continues till 5 h after high water. The returning current is similarly delayed.

The influence of the tidal currents is still felt in the Little Bitter Lake and the tidal friction reduces the current velocities by more than a quarter at Kabret (Grace, 1931; after Miller and Munns, 1974). Gruvel (1936) gives interesting data on the tidal fluctuations in the Bitter Lake. The high tide reaches the Deversoir at the northern end of the Lake two to three hours after the tide at Suez. The tidal span is, according to Gruvel, of 55 cm at Kabret, 55 cm at the Deversoir and 70 cm at Genefe in the South-Western corner of the Great Bitter Lake. This last value is attributed by Gruvel to a concomitant piling up of the waters by the Northeast winds.

With the exception of the small tidal influence, little is known about the current system in the Bitter Lake. Fox (1926) mentions several cases in which changes in the direction of the wind caused inversion of the current in the Lake. Miller and Munns (1974) suggest the presence of a counter-clockwise current in the Lake. In their words "a northerly flow on the eastern side and a southerly flow on the west seems to be borne out by the distortions of the salinity distributions". For the sake of later discussion, it should be mentioned that no author ever mentioned the existence of a throughgoing current in the Bitter Lake.

There is no precise knowledge as to what happens in the stretch of Canal between the Bitter Lake and Lake Timsah. Usually it is assumed that here already the 9–10-months' regime of the northward currents prevails. However, Wüst (1935) indicates an inverse current during July–October. In December 1973 Hecht (personal communication) measured north of the Deversoir a wind-driven superficial southward current, a deeper and dominant northward current and along the bottom, again, a southward flow.

Lake Timsah too, has no throughgoing currents and the water movements are probably mainly wind-driven.

The currents in the northern stretch of the Canal have been much discussed. A more detailed presentation is, therefore, necessary.

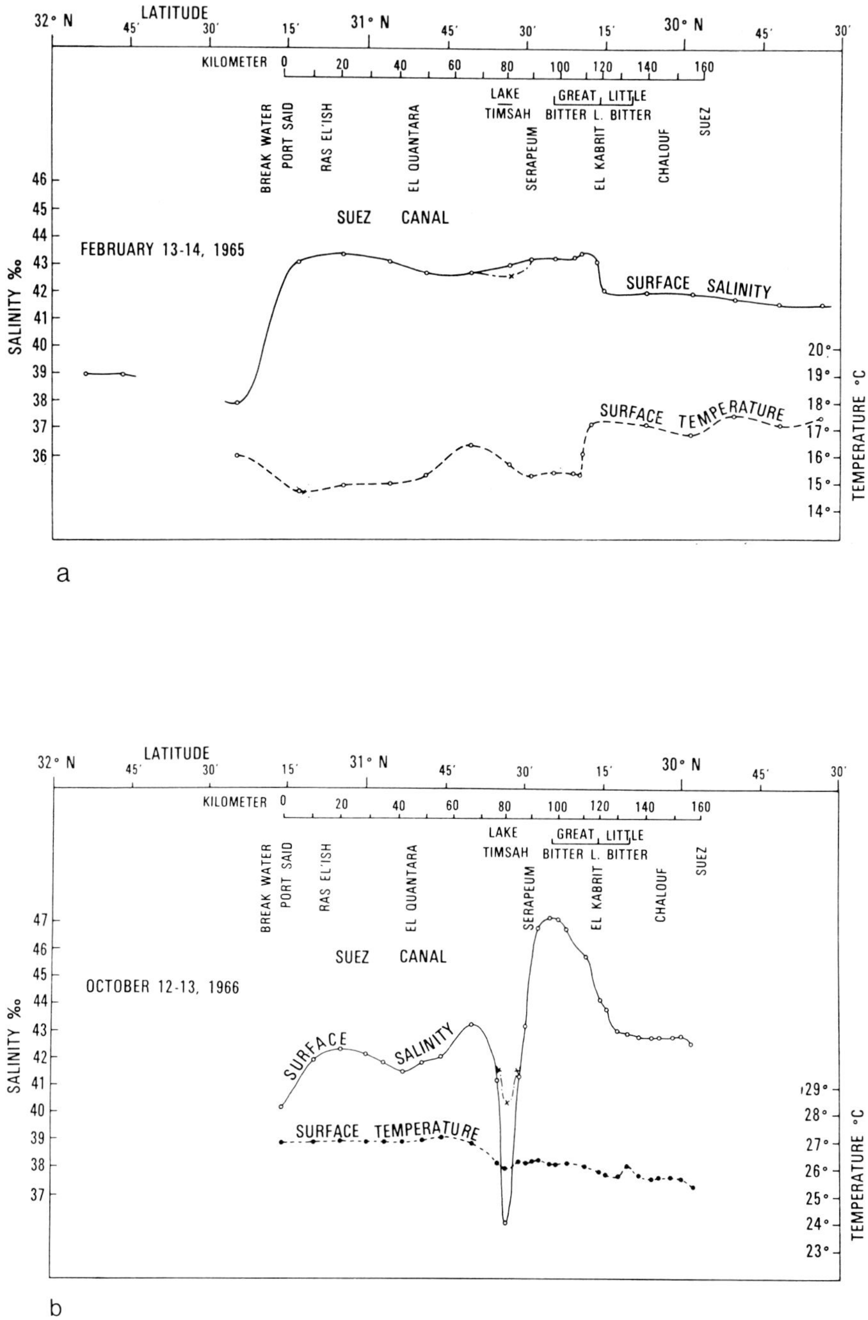

Fig. 22. (a) Salinity and temperature profile in Suez Canal, February 13–14, 1965; (b) salinity and temperature distribution in Suez Canal, October 12–13, 1966 (from Miller and Munns, 1974)

First of all, the tidal differences, an eventual driving force, are minimal in the Eastern Mediterranean. Gohar (1954) indicates a tidal range of 15–40 cm at Port Said. Fox (1926) gives a maximum spring tide value of 40 cm. It follows that there are practically no tidal currents in the northern Canal section. Fox indicates tide-gauge amplitudes of 3.0–3.5 cm at El Qantara (Km 46). In July 1963, during the transit of the "Atlantis II" through the Canal, the influence of the Mediterranean tide extended as far as Ras el Esh—about 30 km (Miller and Munns, 1974; Fig. 22).

The current in the northern Canal is, therefore, influenced by the seasonal water level differences. Starting with Tillier, all the authors concerned agree that there is a prevailing northward current in the northern Canal for three quarters of the year, between October and June. For this, current velocities of 0.5 to 2 knots are given. In June 1967, the current was so strong that SCUBA divers working at El Qantara had to attach themselves to the piers. In November–December 1967 the current flowed a constant north, despite the permanent strong winds in the opposite direction (Ben Yami, personal communication).

It is curious that the older authors and especially Lesseps (1876) made different observations. He wrote: "L'évaporation étant plus active dans le centre de l'isthme qu'aux deux entrées de Suez et de Port Said, le courant vient presque toujours du sud au nord à partir de Suez jusqu'aux Lacs Amers et du nord au sud à partir de Port Said". Keller (1882) even indicates a speed of 0.3 cm/sec for this southward current in the northern Canal. We have to consider that Keller visited the Canal in January, i.e. in the season in which the current is reversed. That Keller was keen in observing the currents is evident from the very accurate description he gives for the diurnal changes in the tidal currents of the southern Canal.

It would be interesting to suppose that the current regime in the Canal changed after the first few decades—eventually owing to the deepening of the Canal.[1]

From July throughout September there is a southward current in the northern Canal. Gruvel (1936) is the only author who mentions that this current starts already in June. During these summer months the level of the Mediterranean waters becomes equal to those of the Gulf of Suez. It is reasonable to assume that under such circumstances the wind, permanently northerly, becomes the driving force of the current. The equalization of the sea levels is due to the loss by evaporation in the summer heat of the Gulf of Suez. This is however, even much more severe in the Bitter Lake, which according to Wüst (1935) lies some 10 cm below the level of the Gulf of Suez.

Morcos (1960) reports on continuous current measurements at two points—south and north of the Bitter Lakes during 1933–35. However, according to Morcos (discussion with Miller and Munns, 1974) there is "a net flow towards the north in the summer and a net flow towards the south in winter". The winter flow

[1] *Note added in proof:* More recently Morcos and Messieh (1973a) and Morcos and Gerges (1974) following Baussan (1938) speak of a transitional stage in the currents in the summer, when water flows both from the Mediterranean and from Red Sea *into* the Bitter Lakes.

lasts 9–10 months and the summer flow 2–3 months. It is hard to reach any conclusions from these data.

Miller and Munns (1974) make some very interesting calculations bearing on the evaporative losses in the Bitter Lake during the months of July and August. Although these calculations are based on a combination of old information obtained by these authors from Morcos with measurements made by the "Atlantis II" in July 1963, the results are interesting. According to these, the Bitter Lake received an excess of seawater inflow from the Red Sea of 432,734 m^3/day. Taking into account the incoming salinity value of the sea water and salinity existing in the Bitter Lake (on the assumption that all the concentration resulted from evaporation) the authors reached a figure of 771,000 m^3/day to be evaporated. To make up for the differences which resulted, "contributions from the Mediterranean of Nile indrafts" are necessary in these summer months.

Many authors, among them Morcos, saw in the summer southward current also a result of the piling up of Nile waters at Port Said, following the opening of the Nile dams which used to occur in these months. After the final closure of the Aswan Dam, Morcos (1967a) mentions that in September 1966 a weak northward current appeared in the Canal.

According to the "Red Sea and Gulf of Aden Pilot" (1955) there is no sufficient evidence to prove that the southward current in the northern Canal was caused by the seasonal rise of the Nile. El Sabh (1968) indeed considers that Morcos' data from 1966 represents a meteorologically abnormal situation and that no fundamental change in the current patterns occurred after the cessation of the Nile flow in early 1966. Oren (1970) agrees that the Nile rise did not play a crucial role.

Even though the surface current flows southward during the late summer, both Wüst (1935) and Gohar (1954) speak of a bottom outflow of heavy saline Bitter Lake water on the bottom of the northern Canal even in the summer months.

Temperatures. There are few relevant data on water-temperature for the Suez Canal. Nonetheless, temperature has often been mentioned in speculations on the animal migration through the Canal.

Without supplying the appropriate data, Wüst (1934) asserted that temperature differences in the Canal system were very small. Fox (1926) and later W. Steinitz (1929) speak of the high water temperatures of the Canal as a barrier to animal migration. Other authors see the Canal merely as a gradient of decreasing temperatures from the Red Sea to the Mediterranean.

The temperature ranges as they appear below result from few measurements: those of the "Pola" in October 1895 and May 1896 (Luksch, 1898); Wimpenny (1930) between 1928 and 1929; Gruvel (1936) in April 1933 and between August and October 1934; El Sabh (1969) in 1966; the temperatures measured by "Atlantis II" and "Chain" in July 1963, February 1965 and October and November 1966 (Miller and Munns, 1974); finally the temperatures measured in 1967 and 1969 by the Hebrew University team (Por and Ferber, 1972).

The maximum temperatures found are around 30° C: in September 1934 30° C at Ismailia; 30.5° C at Kabrit in July 1963; 29.92° C in the Bitter Lake in September 1966; 29.60° C in July 1928 in Lake Timsah; 29.70° C in June 1967 in the Bitter Lake. The high temperature period in the Canal system comprises, therefore, the months June–September.

In July 1963 "Atlantis II" measured 26° C and 26.5° C at the Mediterranean and Red Sea entrances, respectively, whereas in the Canal the maximum was 30.5° C. This is a difference which has to be reckoned with, although it stands to reason that in the shallow lagoons of Suez and even in those near Port Said the temperatures were at the same time probably no less than at Kabrit!

Minimum temperatures in the Canal system are rather low: Wimpenny reports 14–14.3° C in January–February 1929 in the northern part of the Canal and Lake Timsah; surface temperatures between 14–15° C are indicated for the northern part of the Canal for February 1965; and somewhat higher 16° C was measured by Gruvel in April 1933 in the Bitter Lake.

From the open waters of the Gulf of Suez, minimum temperatures of 16° C were reported (Morcos, 1970). However in the shallow lagoons of the northern Red Sea the low winter temperatures may often descend to as low as 10–11° C. In the inshore waters of the Mediterranean, minima of 15.4° C have been measured. Even lower values have been found in the Sirbonic Lagoon (Bardawil Lagoon).

Being exposed to a similar regime of evaporation and winds during the winter, it seems that there is no significant difference in the minimum temperatures of the shallow waters around the Sinai Peninsula. Temperature barriers, both low and high, for the spread of marine organisms should not be taken into account for the

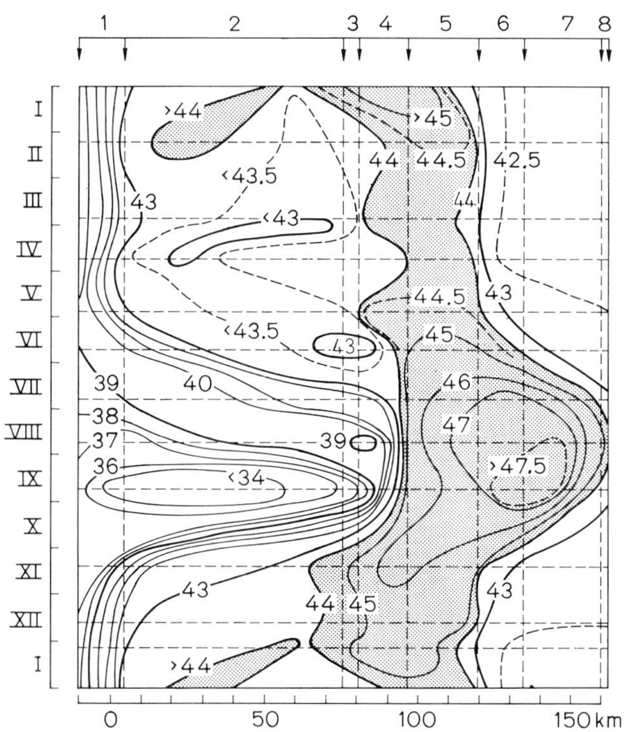

Fig. 23. Isohalines in Suez Canal, from Port Said to Gulf of Suez, at depth of 6 m (from Wüst, 1934)

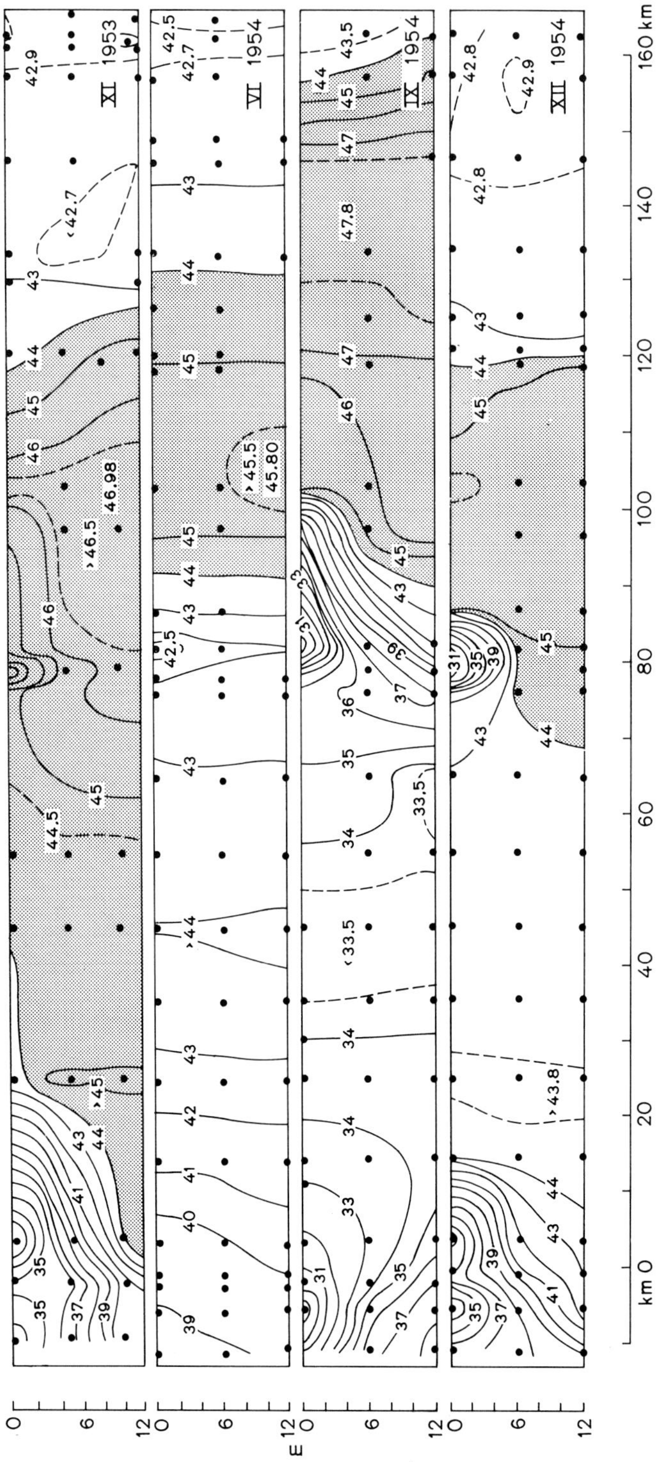

Fig. 24. Seasonal distribution of salinity in Suez Canal (1953–1954) from Port Said to Gulf of Suez (from Morcos, 1960)

Canal and its shallow approaches. A limiting temperature only starts to act outside the Canal in the open Red Sea or along the Levant Coast of the Mediterranean.

Salinity (Figs. 23 and 24). Salinity is probably the most important environmental factor influencing the settlement of organisms in the Suez Canal. The salt content of the Canal waters has been reasonably well investigated during the past century and it soon became clear that it does not result from a mere gradual mixing of the Red Sea and Mediterranean water. The reasons for that are threefold: the dissolution of the fossil salt dome at the bottom of the Great Bitter Lake; the evaporative concentration in the different lakes and shallows; and an inflow of Nile waters at different points of the northern Canal.

The salinity pattern of the Canal system should, therefore, be treated separately for the different sections of the Canal, and we shall start from south to north. Understandably, since the Suez Canal is not the central subject of this book, the presentation below will be restricted to the data relevant to our subject.

The unusual salinities measured in the open sea at Suez—maximum of 44.3‰ in September 1897—have been attributed by Luksch (1898) and Morcos (1970) to an outflow of brine from the Suez Canal. However, Robinet and Lefort (1874) indicate an even higher salinity of 45.38‰ at Suez in December 1864, i.e. before the opening of the Suez Canal. This is a normal salinity in the shallows of the northern Red Sea and the Gulf of Suez. Por (1972) found values of 46‰ at Ras el Misalla (about 15 km south of Suez) and 43.39‰ at Ras Sudr (50 km south of Suez) in 1970. The high salinity values found in the Gulf of Suez are due to the water losses by evaporation in this shallow gulf. According to Miller and Munns (1974), the Gulf is "more conducive to evaporation than any other maritime area" and its high salinity has therefore not to be attributed "to the leaching of the Bitter Lake salt beds".[1]

The Bitter Lakes constitute the main high-salinity barrier in the Canal system.

Early authors, such as Krukenberg (1888a), believed that evaporation is the main factor determining the high salinity of the Lakes. Keller (1882) attached equal importance to evaporation and dissolution of the Bitter Lake salt deposits. Starting with Tillier (1901), who, however, does not give reliable original measurements, attention began to focus exclusively on the dissolution of the salt dome. Based on the original thickness of the dome (13 m as given by de Lesseps) Fox (1929) calculated that all the salt will have been dissolved by the end of the 20th century. Wüst (1935) found a direct relation between the increase in the depths of the Bitter Lakes, i.e. in his view between the progressive dissolution of the salt bank, and the decrease in salinity. In 1951 however, Wüst reached a new conclusion, namely that the decrease in salinity is asymptotical. Krauss (1958) was the first to accept that the salinity of the Bitter Lakes will always remain above normal sea salinity even after complete dissolution of the salt, owing to evaporation. Despite this, Oren (1969) still believed that the final values will eventually be as low as 41‰.

---

[1] *Note added in proof:* Morcos and Messieh (1973b) seem to find evidence that after the cessation of the Nile floods the salinity in Suez Bay may decrease since less Bitter Lakes waters reach this bay.

Since the late fifties there has been no further decrease in the salinity of the Bitter Lakes. Morcos (1960) was the first to accept this fact. Oren (1970) considers that the minimum salinity was reached in 1958.

Another contradiction is evident when Morcos (1967a) writes of salts which are "gradually being dissolved". In fact, in the same paper, Morcos gives analyses from the salt dome collected in 1955. Miller and Munns (1974) report that core and grab samples taken in the Great Bitter Lake in 1965 by "Atlantis II" showed that the salt deposits were probably exhausted. Mud and gypsum crystals were the main constituents of the cores". Etam (personal communication) who sampled cores in thė Bitter Lake in December 1973 found only gypsum and no trace of crystalline salt.

It is therefore evident that since the early sixties the salinity of the Bitter Lakes has been entirely determined by evaporation. Por (1971 b) stated that the salinity of the Bitter Lakes will always remain above that of the open sea. Miller and Munns (1974), reporting salinities between 43‰ and 48‰, consider the present salinity of the Lakes as being a seasonal phenomenon with a maximum in summer and a minimum in winter.

Muromtsev (1960, 1962) measured salinities in September 1959 and in July 1960. In the Bitter Lake he found 45.98–46.87‰, and 44.79–46.06‰ respectively. The maximum salinity value found by this author was 48.06‰ at Km 155, "caused by strong evaporation". The most recent measurements give 46.17‰ for January 1969 (Por and Ferber, 1972) and 45‰ for December 1973 (Etam, unpublished).

To give an idea of the magnitude of the evaporation in this area, Miller and Munns (1974) indicate values (seasonal?) between 3.9 to 6.0 mm/day. This would add up to an average of 1.8–1.9 m/year.

It is evident therefore, that the Bitter Lakes will behave in the future as one of the littoral lagoons of the Red Sea, with salinity fluctuating seasonally around the value 45‰.

There has always been a seasonal fluctuation in the salinities and it seems that this even decreased over the years: it was 8‰ in 1924 (Wüst, 1935) and in 1954–55 about 5.5‰ (Morcos, 1967b). Presently I would suggest an amplitude of below 5‰.

It is hard to follow the actual decrease of salinities in the Bitter Lakes, however an attempt to show the data graphically is given in Figure 28.

A last point to be made about the Bitter Lakes' salinities is the tendency towards a saline stratification. This was more evident several decades ago and although still to be observed, shows at present only a narrow range of differences. A difference between surface and bottom values of about 6‰ seems to have existed in 1872 (Durand-Claye, 1874). In October 1924 the difference was less than 4‰ (Fox, 1926); however, in January–March 1925 it increased to some 8‰ (Wüst, 1935) (Fig. 23). For 1954–55 Morcos (1960) gives less than 1.5‰ (Fig. 25). However in 1966, El Sabh (1968) again gives a value of 2‰. In the salinity profiles of Miller and Munns (1974) differences of 1.0–1.5‰ between surface and bottom appear.

Of special interest are the hypersaline lagoons and pools which border the Bitter Lakes, especially along the eastern shores. The largest of these, in the northeast corner of the Great Bitter Lake, had a salinity of 91.08‰ in January

1969 (Por and Ferber, 1972). Fox (1926) mentions seven brine pools near Kabret—within about 200 m from the shore. The pools had a wide range of salinities, from 52‰ to near saturation (density 1176). All these lagoons and pools are formed by spilling–over of the lake waters during the winter storms. Their highly saline water, as well as that of the surrounding salt wetlands, probably contributes a sizeable share to the increased salinity of the Bitter Lakes, either through direct leaching out by waves or through percolation in the sediment.

Lake Timsah is characterized by a marked saline stratification. It receives salty bottom waters from the south and these are overlain by a layer of diluted water coming from the Nile. The fresh water comes from the Nile through a canal which supplies drinking and irrigation water.

Until 1891 this canal flowed out freely into Lake Timsah and therefore during the Nile floods huge amounts of fresh water reached the lake. Tillier (1902) even considers that some preexistent marine species were exterminated by these floods. After 1891 the fresh-water canal was extended northward to Port Said. However, even after this date the waters of the fresh-water canal ("Abassa Canal") still reached the lake either from the sewers of the town of Ismailiya or as agricultural runoff from the lush fields of the western bank.[1]

Since Lake Timsah has many shallow lateral lagoons, these carry almost exclusively fresh-to-brackish waters to a depth of 30 cm (Fox, 1926). While the deep center of the Lake is occupied by a saline bottom layer, the stratification sometimes reaches a considerable amplitude. In December 1924 Fox reports a difference of 9‰ between surface and bottom. W. Steinitz (1929) indicates that at the surface the salinity may fluctuate between 7.86‰ to 47.6‰. Morcos (1960) found less than 30‰ at the surface and 45‰ at the bottom. In June 1967, surface salinity in the lake opposite Isma'iliya was 29.3‰ (Por and Ferber, 1972). El Sabh (1969) gives surface salinities of 35‰. Miller and Munns (1974) report strong differences between surface and bottom: in October 1966 from 36‰ to 41‰, and in November 1966 from 37‰ to 42‰. They speak of a "strong intrusion of fresh water" and "extra volumes of Nile water entering at Lake Timsah". In February 1965 Lake Timsah was almost homogenous and of the same salinity as the Bitter Lake, i.e. $\pm$43‰.

Thus the stratification of Lake Timsah is still considerable and is a seasonal phenomenon with a peak in the late summer.

The northern stretch of the Canal between Lake Timsah and the Mediterranean is exposed to an inflow of low salinity Mediterranean water from August–October. The inflow is mainly caused by the equalization of levels of the two seas—or even a decrease of the Red Sea level below that of the Mediterranean, in late summer (see above). Before the building of the Aswan High Dam, this period also coincided with the seasonal opening of the Nile barrages and the flow of large amounts of fresh water into the sea not far from Port Said.

The salinity decrease in the northern Canal was first observed in the salinity data of Durand-Claye in 1872. In the harbor of Port Said the salinity was 23.1‰ while at Km 60 it was 45.5‰. In the rest of the Canal the concomitant salinities

[1] *Note added in proof:* Morcos (1975) puts the influx of agricultural freshwater runoff at 210 million cubic meters per year and expects this influx to increase in the near future.

were above 60‰. From these data we may deduce that the samples were collected in the August–October period; this is however not by any means certain. In October 1895 the "Pola" found no decrease in salinity and at Km 60 the value was 51.2‰ (Natterer, 1898). It can eventually be assumed that in the first period of Canal history, the salinity was so high that the diluting influence was not felt so much, or conversely that the current regime in the northern Canal was then different (see above).

Tillier (1902) is again the first author to mention the connection between the summer current regime and the appearance of low salinity waters in the northern Canal. Wüst (1934) gave the classical presentation of the phenomenon.

Morcos (1960) reports for September 1954 33.15‰ at the Canal entrance and 36.29‰ near Lake Timsah. The respective values for September 1964 were 32.25‰ and 34.83‰. These are, however, surface values since in September 1964 at the Canal entrance the salinity at only 5 m depth was already 39.02‰.

After the cessation of the Nile flow through the closure of the Aswan Dam, in October 1966 the "Chain" measured salinities between 40–43‰ in the northern Canal, whereas in the Bitter Lakes it was 47‰. It seems therefore that in the late summer period there is no marked decrease in salinity in the northern Canal.

Several authors assumed that the late summer period with the southward current and lowered salinity is—or rather was—especially propitious for the influx of Mediterranean species into the Canal. However, in the season when salinity is at its lowest in the northern Canal, maximum seasonal salinities are found in the Bitter Lake area. In September 1954, for instance, the salinity range from north to south reached 14.6‰ and in September 1964 even 15.6‰. On the other hand in April of the respective years the salinity variations along the whole Canal system were restricted to only 2‰.

One may eventually assume that after the closure of the Aswan Dam in 1966 some advantage resulted for the southward migrating animals since in the period in which a southward current moves into the Canal, there are no more of the sharp changes in salinity which the organisms had to face before.

Pollution. Little is known about pollution in this highly navigated waterway. One can accept a priori that the Canal system suffered much from pollution and that the whole environment was comparable to that of a big harbor. There are no data concerning the restricting influence of pollution on the Canal biota.

The only data found in literature are those of Fonselius (1966, 1967) who found unusually high values of zinc and copper in the Bitter Lake. Strangely, Fonselius attributed this to some sort of influence from the Northern Red Sea, in which the values of these ions were also high. Morcos (1970) however, rightly assumes that these concentrations may be due to pollution.

Oil pollution due to leaks and bilges is probably high. In the littoral of the Bitter Lake near Deversoir in 1969, I saw considerable amounts of oil and tar mixed with the beachrock and the sandy conglomerate (Por, 1971b). Considerable oil pollution is reported from Suez where big oil refineries are situated west of the harbor of Port Ibrahim. Pollution there was already considerable in the thirties when Gruvel (1934) considered the area as "literalement empoisonnée par le pétrole, le mazout ou leurs déchêts de fabrication".

Pollution, chiefly from bunker oil, may also explain the presence of the much-mentioned "gelatinous mud" on the bottom of the shipping channel through the Bitter Lakes. The first record of this, along the eastern coast of the Lake, is to be found in Tillier (1902) who speaks of a "vase noire remarquablement molle et collante que nous avons pu parfois comparer à un épais sirop". Fox (1926) speaks of a "black evil-smelling gelatinous mud". Thorson was very interested in this gelatinous mud, which he considered to cover the bottom over great distances [see correspondence between Steinitz and Thorson Israel J. Zool. **21** (3–4), pp. 143—147, 1972]. Beets (1953) found that the foul bottom layer may be due to "stagnancy and exhaust products of mechanically driven ships". As late as 9 July 1974 the *International Herald Tribune* reports "a layer saturated with oil and salt... fuel which leaked from ships" which was encountered by American naval divers who worked in the clearing operation of the Suez Canal.

It seems that the mysterious "gelatinous mud", one of the curious features of the Canal, is a bunker oil, salt and water mixture covering parts of the bottom of the Bitter Lakes.

2.4 The Bottom of the Suez Canal

The Suez Canal is a habitat for many species—regardless of where they originate or their chances of becoming invaders of the opposite sea. For most of these animals the types of bottom found in the Suez Canal are of crucial importance. Conditions are more or less natural and undisturbed in the Bitter Lakes and Lake Timsah. In the Canal proper, the continuous dredging and the periodical deepening and widening are very important as stability factors. The waves and turbulence caused by the many ships passing the Canal day and night are a further factor to be considered as deeply influencing the sediments of the Canal sections.

Most of the Canal used to be dredged once every three to four years. The minimum frequency of the dredgings was five–six years, maximum two years. W. Steinitz (1929) gives an average time span of 2–3 years between dredgings. To that one has to add the several widening and deepening projects which have been carried out over the past century. As a rule, the Canal was widened along its east bank. The enlarging of the Canal did not result in more undisturbed conditions along the banks and the bottom, since the projects were always carried out as an answer to the increasing size and draught of the throughgoing ships.

Following Gruvel's detailed description (1936) the Canal bottom can be characterized in the following way: in the North to about Km 55 the bottom is sandy, mixed with black mud of lagoon origin. From Ballah onwards the bottom becomes "cleaner", mixed with gypsum concretions. Crossing the ridge of Guisr (Km 72) the bottom is marly with some rocky outcrops. Sandy mud predominates between the two lakes (Lake Timsah and Bitter Lake). South of the Bitter Lakes the sand is cleaner, with a mixture of gravel and shell. In the stretch crossing the ridge of Shallufa (Km 140), rocky outcrops are again found on the bottom.

With the exception of the northern stretch, the bottom is clean enough to allow for a dense population of the lancelet *Branchiostoma lanceolatum* in the sediment. This has been a stable and frequent inhabitant of the Canal stretches for at least the last 50 years, from the Cambridge Expedition of 1924 till 1972, when it still was widespread. The lancelet needs fairly clean sandy bottoms, so that the much-repeated talk about the progressive silting of the canal bottom is not supported by facts. Gohar (1954) speaks, for instance, of the fact that "during the first years after the opening of the Canal the deposition of mud had not gone so far as to impede the settlement of plants and animals ... Thus it is highly probable that, during those years, the passage of the faunas ... was more active than it is now". There are, however, no indications that the migration was more active in the first years after the opening of the Canal.

There are likewise no data regarding a gradual silting up of the canal bottom. As early as 1858 Schleiden expressed fears that an artificial canal leading through the desert would soon be shoaled up by sands, as in the Sirbonic Lagoon. In the different war periods in which the Canal was not subjected to maintenance and especially during the last seven years of inactivity (1967–1974), fears were expressed that large stretches would have to be literally dug out again from under sand. This, however, was not the case for the several months of inactivity in 1956 [see Thorson in his letter exchange with Steinitz (1972)]. No dredges were used to open the Canal in 1974 and the only problem mentioned as hindering the opening to navigation were the mines, other war remnants and ship wrecks.

It is apparent that the currents in the Canal stretches are strong enough to sweep away mud and sand deposits. This does not, however, mean that the Canal bottom is a suitable substrate for a slowly developing population of sedentary animals. *Branchiostoma* is the best example of an animal which lives buried in a current-swept substrate, capable of adjusting itself quickly and by adroit movements to the shifting of the substrate.

Most of the Canal banks are plated with stone and concrete slabs. The covering with stone started near the end of the 19th Century and is mentioned by Tillier (1902). To this, one must add the ever-increasing surfaces of piers, pontoons, buoys and, for some periods, ships wrecked or sunken in the Canal. Therefore, at the surface of the Canal there is no dearth of hard substrate for a few species of hardy intertidal organisms to settle. This circumstance was overseen in one of my previous papers (Por, 1971b), where I stated that hard substrates are rare along the Canal. The stone banks are settled by some very hardy intertidal organisms. However, in Lake Timsah and the Bitter Lakes, the shores are sandy, with only restricted hard bottoms of beach-rock type.

If the lakes contain nearly 90% of the waters found in the canal, they offer also the great majority of the bottoms available to the organisms entering the Canal. Most of the lake bottom is natural and has never been dredged out. Therefore, soon after the flooding, normal sedimentological processes started and similarly also an undisturbed buildup of benthic communities began. The fact that the bottoms of the lakes are almost exclusively sedimentary is of crucial importance for the type of population which inhabits the Canal.

Our knowledge of the bottom of Lake Timsah (Fig. 25) is based on Tillier (1902), Fox (1926), Gruvel (1936), and on Tortonese (1952). According to Tillier only some 75 ha have been dredged and dug out. To this one must add, according to the same author, dumping places of out-dredged material, "bancs de vidage", which at that time covered some 60 ha. Those dumping places have no doubt

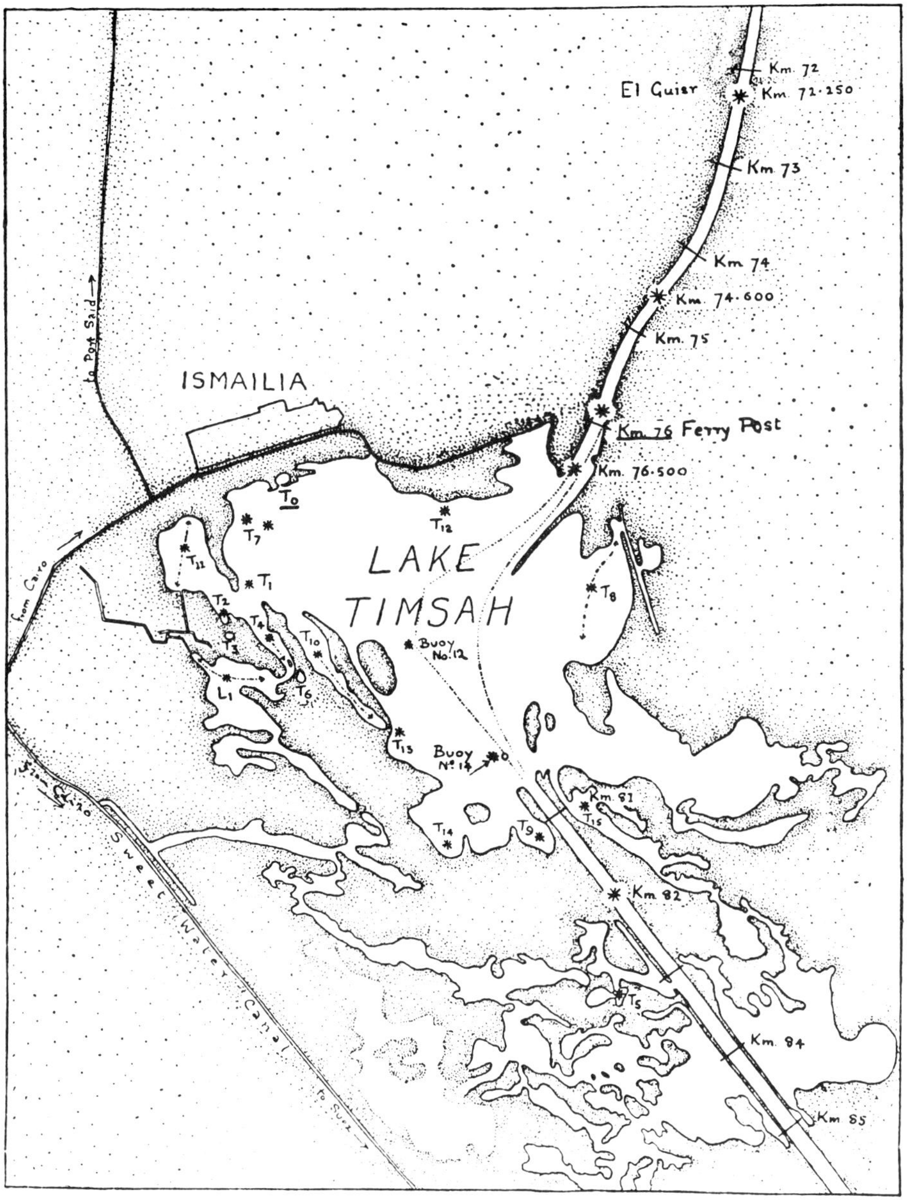

Fig. 25. Map of Lake Timsah (from Fox, 1926)

increased in number and surface over the years: their suitability for settling by benthic animals has to be considered as intermediate.

Black mud, smelly and gelatinous, "very like that of the Great Bitter Lake" (Fox, 1926) appears in the north-west corner of Lake Timsah at 7–8 m depth. Gruvel (1936) reports the presence of black mud "plus ou moins liquide et parfois fétide" in the middle of the lake. According to this author only *Murex tribulus* lives on these muddy bottoms. On the other hand, Fox reports grey sandy mud from the center of the lake with abundant ophiuroids and leucosiid crabs. Black mud with ophiuroids appears also at another station in the middle of the lake. It appears thus that the dredged bottoms of Lake Timsah present difficult conditions for benthic animals, though still suitable for some species.

The natural undredged bottoms of the lake, at an average of 6 m depth, as well as the many shallow lagoons are covered with mud at depths exceeding 2–3 m; however, this mud is already rich in fauna. These bottoms are characterized as grey muds by Gruvel.

Gradually this oxygenated mud is covered with a layer of sand and around 2–5 m meadows of the flowering plant *Halophila stipulacea* appear. The shallows of the lagoons are pure sand (Tortonese, 1952)—the lagoons of the western shore are already covered with fresh water and overgrown with *Ceratophyllum*. In the eastern lagoons *Halophila* is found even on the shallowest sands.

Lake Timsah is given by various authors as a textbook case (following Fox, 1926) for the presence of oxygen on the bottom despite saline stratification. Since the water is clear, water plants and algae provide abundant oxygen supply, probably down to some 3–4 m depth.

The bottoms of the Bitter Lakes are much better known. Gruvel (1936) (Fig. 26) and Beets (1953) give bathymetric and sediment maps of the Bitter Lakes. A recent transect sampled in 1973 by Y. Etam (personal communication) also gives interesting information.

The problem of the so-called "gelatinous mud" and the gradual dissolution of the salt bank have already been discussed in a previous chapter.

Here is the relevant place to present the description of the basin of the Bitter Lakes before the flooding by the Canal waters, as given by Aillaud (1868). The basin was a typical "sabkha": a salt marsh like many others around the Red Sea and Persian Gulf and the Sabkhat el Bardawil (mentioned above as the Sirbonic Lagoon) (Fig. 27). Aillaud speaks of the thickets of *Tamarix* on the east shore as being inhabited by a rich avifauna counting among others such water fowl as pelicans, flamingoes, ducks, and geese. These are frequent on the sandy shoals which emerge from the water-filled depression which surrounds the salt bank. As far as one kilometer from the central salt dome there were big puddles of very clear water. The bottom around was muddy and contained many mollusc shells.

At the time of Tillier, the Bitter Lakes were filled up and had sandy shallows around the central deep. Fox's samples came mostly from the Little Bitter Lake: samples of black anoxic mud, one sample of grey muddy sand at 11–12 m depth with a fairly rich fauna, and two shallow samples from bottoms covered with the sea-grasses *Halophila* or *Diplanthera (Halodule)*.

Gruvel indicates that the whole central part of the Great Bitter Lake is covered by a gypsum-admixed mud. This is an anaerobic bottom: a "zone morte" where only a few specimens of *Murex tribulus* and *Gastrochaena* live.

Beets (1953) did not investigate the deepest bottoms of ± 15 m. However, from the central shipping lane he reports at 10.80 m and 11.30 m depth, clay or sandy clay devoid of vegetation. The bottoms below 9 m are, according to this author, fouled by a combination of stagnancy and the exhaust products (see also previous chapter). Etam in 1973 (pers. comm.) reports from 10–16 m depth a variety of

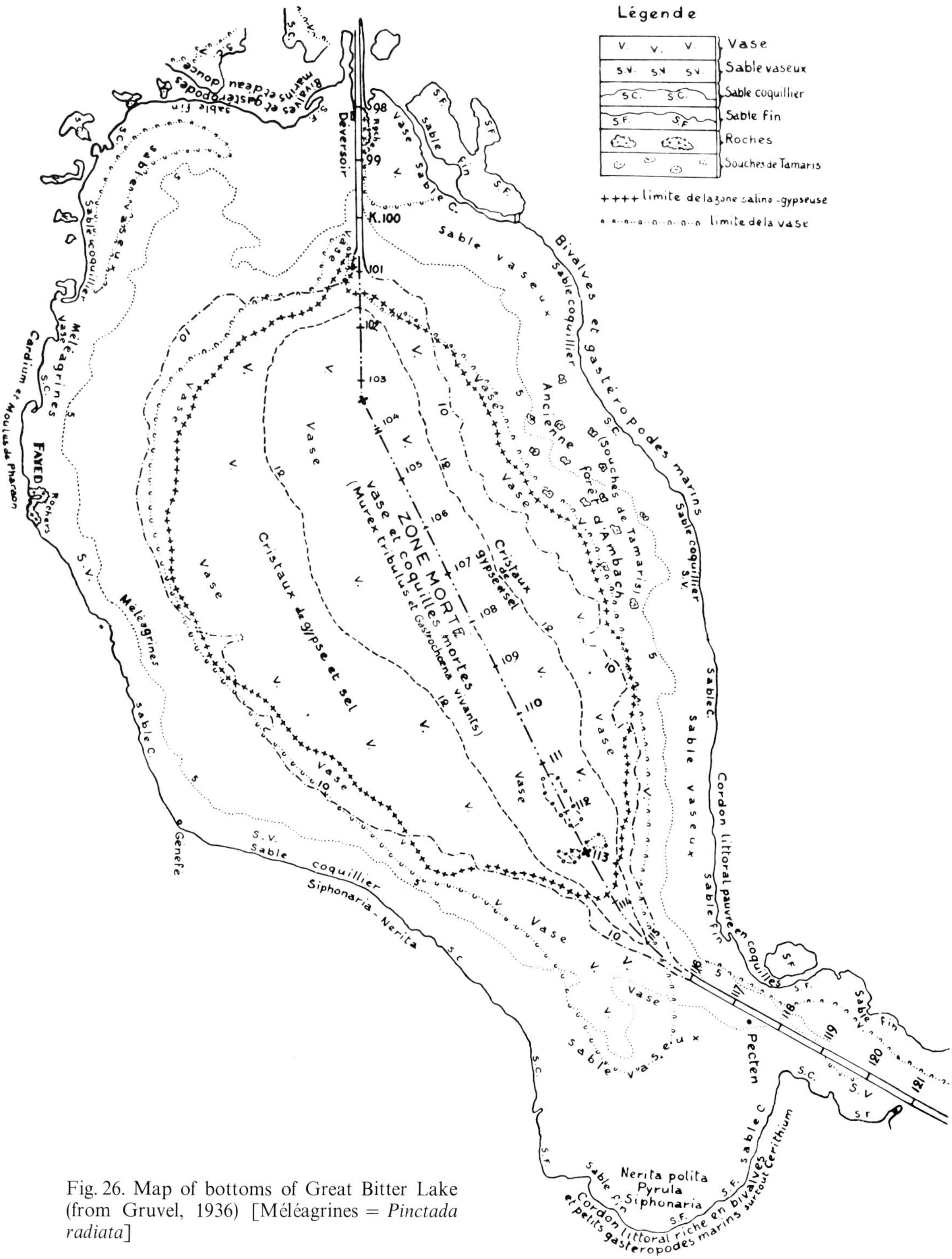

Fig. 26. Map of bottoms of Great Bitter Lake (from Gruvel, 1936) [Méléagrines = *Pinctada radiata*]

Fig. 27. Basin of the Great Bitter Lake (near el Ambach) before flooding by Canal (after a contemporary lithograph, from Gruvel, 1936)

bottoms: a light brown mud, a dark reductive mud, clayey mud mixed with gypsum and sandy muds.

It is evident that the bottoms below 9–10 m have suffered and are perhaps still suffering locally from periods of lack of oxygen. This may be due in the first instance to conditions of saline stratification in the Bitter Lake. Even under the present circumstances of relatively lowered and homogenous salinities, pycnoclines hindering circulation can still occur locally and temporarily.

Older information regarding the shallow bottoms of the lake, the "zone vivante" of Gruvel, is more than patchy. According to Tillier the eastern shores are muddy, covered with sand and overgrown with very scattered patches of algae. The Little Bitter Lake has, according to this author, bottoms of much harder consistency. Fox (1926) gives only little information about the shallow bottoms of the Lakes. He indicates the presence of *Halophila*-covered sand bottoms between 1–3 m depth. Gruvel (1936) gives a good map of the Great Bitter Lake and its bottoms. According to him the shallowest are shell-admixed sands and around 5 m depth there are muddy sands. Gruvel states that it is rare to find living specimes of animals and plants below 6–7 m depth, although the sandy muds descend to 8–9 m depth before being replaced by the gypsiferous muds (see Fig. 26).

The most complete information is supplied by Beets (1953) who dredged the bottoms of the Great Bitter Lake. His 47 stations cover the whole lake with a most satisfactory grid. According to Beets, there are three types of "Plant Zones" characterizing the shallow bottoms down to 10 m depth, the depth of the "fouled" bottoms. Plant Zone 2 is, according to Beets, covered with "*Sargassum* prob.*crispum*". In fact, there are probably three species of dominant algae there: *Sargassum dentifolium*, *S. subrepandum* as well as *Cystoseira myrica* (Lipkin, 1972a). The upper limit of the *Sargassum* zone is about 2–4 m. This type of algal association was found by Beets along the western and most probably the southern and southeastern shore. In Beets' opinion there is a connection between the presence of distinct plant zones along the western and southern shores and the prevailing northerly winds.

Plant Zone 1 of Beets occupies the shallows of the western and northern shores. It seems to this author probable that the zone is present with some interruptions all around the Great Bitter Lake. The leading plant in this zone is the sea-grass *Halophila stipulacea*, almost exclusive in Beets' words. According to Lipkin (1972a) *Halophila* is intermingled to some extent with another sea-grass, *Halodule uninervis* and with the alga *Caulerpa racemosa*. Plant Zone 1 extends from just below water level [usually 25 cm depth, Lipkin (1972b)] down to 2–4 m and the *Sargassum* zone on the western and southern(?) shores. On the eastern shore the *Halophila* meadows form the only subtidal vegetal association.

Beets' Plant Zone 3 is rather undefined and formed of very shallow patchy algal cover. It is probably similar to the patches of *Cladophora* species growing according to Lipkin on very shallow sands, in fact attached to the protruding shells of *Mactra olorina* and other lamellibranchs and gastropods. The alga *Acetabularia calyculus* appears in the spring, growing on shells admixed to the shallow-water sand (Lipkin, 1972a).

The different plant associations grow on sandy-to-muddy bottoms. The bottom of the eastern shore is much shallower and less steep: here transition from sand to mud is very gradual. The western shore is much steeper, much more wave-battered and therefore the bottom is heterogenous, sometimes with a considerable amount of shell and gravel and locally even with rocky outcrops.

The little we know of the deeper and shallow bottoms of the Little Bitter Lake seems to indicate that there are no radical differences from the Great Bitter Lake: the sediments seem to be coarser and rocky outcrops more frequent.

The shores of the Bitter Lakes are generally sandy however, locally and especially in the northern and northeastern corner, a beachrock–like formation is found. There are also small segments of shore covered with boulders.

The fauna and flora which develop on the boulders will be discussed in the following chapter. The beachrock is probably a relatively new formation. Gruvel (1936) made good photographs of the beachrock layer and the legend calls it "tuf vaseux noir". Por (1971 b) mentions it as a "presently building-up peculiar beachrock". Finally, Lipkin (1972 b) delves into more detail of the genesis of the Bitter Lake beachrock: according to him, blue-green algae of the genus *Lyngbia* cover the sediment with a crust and filaments penetrate between the sediment particles, packing them together. The process is repeated in the opposite direction when new sediment is added and covers the original crust; filaments grow upward through the new sediment layer until a new surface crust is established. The process, which results in precipitation of interstitial calcium carbonate, is repeated several times. Por (1971 b) attributes a role to the sponges, ascidians and *Brachidontes* byssus in the cementation of sediment, as well as to bituminous residues of oil. The characteristic alga growing on the littoral hard bottoms is *Digenea simplex* (Lipkin, 1972 b).

When speaking of the bottoms and shores of the Bitter Lakes, the littoral layers of dead shell and echinoderms must be mentioned. These form thick and extremely rich collecting grounds for those interested in the fauna of the Lakes. These littoral accumulations have been mentioned by all the authors who collected in the Lakes and they are indicative of a very rich animal life.

A few words remain to be said about the bottom of the southern stretch of the Suez Canal, between the Bitter Lakes and Suez harbor. The bottom is still locally muddy but the shores are marly with some rocky outcrops. Gruvel again mentions the presence of the "tuf argileux" which is probably the above-mentioned beach rock. At Shallufa the Canal passes between rocky banks and as one advances further towards Suez, the bottoms become more and more rocky. In this last stretch of the Canal, there is, according to Gruvel, again a rich population of *Branchiostoma lanceolatum*.

2.5 The Biota of the Suez Canal

The Suez Canal should justifiably be considered and investigated as an environment for life per se. Although basically man-made, it has (at least in some stretches), a past history of natural existence and can be considered by every standard as a chain of several lagoon environments. As pointed out by several authors, but most satisfactorily by H. Steinitz (1968) the Suez Canal is not only a pathway but also a habitat. Although at times the Suez Canal was seen only as a funnel, a corridor through which animals pass like ships from one sea to the other, it is evident that the 162 km could not be passed in one attempt, even by swift swimmers. Keller (1882) spoke of an advance of the biota like "caravans" making stop-overs from time to time; Tillier (1902) separates "espèces fixés" among the fish of the Suez Canal; Gruvel (1936) speaks of the necessity for the species to find "zones de stabulation" in the Canal. The authors, and most recently Thorson unpublished, 1971) never completely ruled out the possibility of planktonic larvae being transported through the Canal by currents during one life-cycle step, i.e. being hatched in the open waters of the Red Sea and reaching metamorphosis and settling down in the open Mediterranean. This hypothesis was, however, never supported by any finding in the field.

Lipkin (1972a), when he opposes my "step by step" advance through the Canal (Por, 1971b) and emphasizes the "one-jump" possibility, refers only to passive transport by ship. H. Steinitz (1968) accepts as "rare incidents" only, cases in which one individual passed the whole length of the route from one sea to the other.

Therefore, Lessepsian migration cannot be understood without an analysis of the living world of the Suez Canal which acts as a filtrating funnel. Elsewhere (Por, 1971b, 1973b), I have postulated that the best and perhaps the only proof that an Indo-Pacific species found in the Mediterranean is a Lessepsian migrant, is the fact that it is living in the Suez Canal. Only those animals able to survive in the Suez Canal environments for some periods or generations, eventually become Lessepsian migrants.

The discussion of the biota of the Suez Canal has every right to be considered a subject for a monographic study. I shall confine myself in the following sentences only to those aspects necessary for the understanding of Lessepsian migration, i.e. the zoogeographical end-product of a successful crossing of the Suez Canal.

One more reservation is necessary. Most of the knowledge concerning the living world of the Suez Canal relates to the Bitter Lakes. Since 85% of the water volume of the Suez Canal is contained in the Bitter Lakes, this circumstance is welcome. However, the lack of information, especially concerning Lake Timsah, is very regrettable.

The following discussion will be brought under three main topics: (1) history of colonization; (2) aspects of floral and faunal diversity; (3) zoogeographical analysis of the Canal population.

2.6 History of the Population of the Canal

Nothing is known about the animal and vegetal populations of the very first years of the Canal. Only Tillier (1902) remarks briefly that during a crossing of the Suez Canal in June 1871 (less than two years after its opening) the waters were rich in fish, especially *Sciaena aquila*. Since this fish feeds on mullets, Tillier assumes that Mugilidae were also present at that date.

According to this same author, even the shallow service channel of two to three m in depth built in 1863 southward from Port Said already contained fish.

The first faunal list—a list of molluscs by Fuchs (1878)—is based on collections made in 1876. It was only in the winter of 1881–1882 that the first more intensive general collecting was done by Keller. At this date Lake Timsah was listed with already 23 species of animals, and there is no doubt that this list of Keller is incomplete.

The colonization of the newly dug canal was very rapid. Therefore we can barely hope to reconstruct the process: by 1880 the colonization had already advanced a long way from its beginning.

It can be assumed that the water bodies in the Isthmus of Suez were never devoid of life: some of these species, called "Isthmus fauna" or "third fauna" (Por, 1971a, b) were immediately incorporated into the fauna of the new canal. *Brachidontes variabilis (=Mytilus variabilis, =M. pharaonis)*, a typical Red Sea species, was already reported by Fuchs in 1876 at Port Said. In the same year *Mactra olorina* has been found far to the north beyond El Qantara (Kantara). These shells, together with *Cerastoderma glaucum (= Cardium edule)*, the cerith *Pirenella conica*, *Balanus amphitrite*, possibly some of the sphaeromid isopods and a whole list of benthic copepods (see Por, 1971b) and also *Aphanius dispar* might have been part of an autochthonous euryhaline isthmus fauna existing before the new waterway.

The euryhaline–estuarine type of fish were quick to follow. *Sciaena aquila* was accompanied by the closely related *Umbrina cirrosa* (Keller, 1882). From the list of fishes given by this author several other estuarine species can be listed: *Mugil cephalus (=Mugil oëur)*, *Dicentrarchus lupus (=Labrax lupus)* and *Crenidens crenidens (=C. forskalii)*. Keller was not an ichthyologist and his collecting was not at all intensive. However, his list gives some further information. First, the above-listed macroscopic species of the Isthmus fauna were all found by Keller. To the list must also be added an unidentified nereid polychaete which Keller found "everywhere under stones". Among the molluscs a few other species were in full advance at Keller's time: *Solen vagina*, *Pholas candida*, *Pinctada radiata (=Meleagrina margaritifera)*, *Anatina subrostrata* and *Cerithium scabridum*.

It is interesting to note that the third wave of settlers was still confined in Keller's time to the southern sector of the Canal, barely reaching the Little Bitter Lake. Examples are *Circe pectinata*, *Murex tribulus (=M. crassispina)*, *Fusinus marmoratus*, *Diodora rüppelli (=Fissurella)*, and *Clanculus pharaonis (=Trochus)*. Algal vegetation was scarce in 1882. Ascherson, who visited the Canal sometime around 1875 reports only a "*Cladophora*"-like alga. Keller mentions however the massive advance of *Sargassum* into the southern sector of the Canal. Keller foresaw that the littoral fauna would start its main advance only after the successful development of an algal vegetation in the Canal. Nevertheless in Keller's time the relatively few species were present in huge amounts: in Lake Timsah the stone slabs were "black with countless *Mytilus*" and harbored "thousands of *Sphaeroma*". *Balanus* settled the whole length of the Canal in "unbelievable" numbers. *Pholas* is found north of El Guisr by the thousands, and punctures the banks "like a sieve". In the littoral zone of Lake Timsah "billions of molluscs" live and form whole shell banks.

Krukenberg (1888a) who was in the Canal area in 1886 does not add much to our knowledge. He mentions the presence of the medusa *Cassiopea andromeda* in

Lake Timsah and mentions that many *Penaeus* sp. were fished in Lake Timsah. Holthuis (1956) identifies this species as *Metapenaeus stebbingi*. Krukenberg mentions the first capture (in Lake Timsah) of one specimen of *Portunus pelagicus*. He also supposes that Keller's „kleine Krabbe" from Timsah was probably *Pilumnopeus vauquelini*, an identification accepted by Holthuis (1956).

From all that can be reconstructed from Fischer's, Keller's, and Krukenberg's information during the period 1870–1890, most of the Suez Canal was inhabited by a hyperhaline–estuarine fauna (see Por, 1973c), forming everywhere huge biomasses. The metahaline fauna of the Red Sea was still mostly limited to the southernmost section of the Canal.

A change in the biota of the Suez Canal occurred in the last decade of the 19th century. The maximum salinity decreased to below 60‰ even in the deeper levels of the Bitter Lakes. The significance of this fact was that the metahaline marine species of the Red Sea could settle in the Great Bitter Lake and thereafter extend their range over the whole of the Canal.

From that point onward the predominance of the Red Sea species in the Canal became established and the way for the colonization of the Eastern Mediterranean lay open.

This new situation was exemplified by the advance of the swimming crab *Portunus pelagicus* through the Canal (Fox, 1924, 1926; W. Steinitz, 1929). In 1889 the crab was still mainly confined to the southern segment of the Canal. In 1893 it was caught in great numbers in the Bitter Lakes and in 1898 it had reached Port Said. By 1902 it was commonly caught at Port Said.

Unfortunately, for the whole period extending from Krukenberg's collecting in 1886 till the Cambridge Expedition in 1924, no general collecting was done in the Canal. There is the very thorough study of fish by Tillier (1902), and for molluscs the similarly good works of Bavay (1898), and Tillier and Bavay (1905). However both authors spent several years in the Canal area and the years of the publication of their studies indicate the latest date for the faunal picture they present. There are however, no dates or seasons in which the different species were collected or seen by these authors. In 1902 a collection of Copepods were sampled through the pump of a throughgoing ship by Professor Herdman (see Thompson and Scott, 1903; Por and Marcus, 1972).

As far as we can judge, by the turn of the century the Canal was already inhabited by the same fauna as today. The fact that the Bitter Lakes (and probably also Lake Timsah) have gone through rapid successional stages until reaching a climax lasting from the beginning of this century, has been pointed out by Por (1973c). Ben-Tuvia (1975b) collected in the Great Bitter Lake in 1973 only species of fish which had been already reported by Tillier (1902) 70 years ago. Comparing all the known data on molluscs, Barash and Danin (1972) did not find any radical change since the fairly exhaustive list of Tillier and Bavay (1905).

Regarding the other animal groups one has to rely on the results published after the Cambridge Expedition in 1924, the Gruvel Expedition in the early thirties, the field notes of 1944–1945 by Tortonese (1952) and data collected by Beets in 1950 (Beets, 1953; Holthuis, 1956). If one compares the results of the Hebrew University–Smithsonian Institution collecting in the Suez Canal between 1967–1972 (Por and Ferber, 1972; Por et al., 1972 and others) it is evident that the same species are as widespread and dominant as 50 years ago. Por and Marcus (1972) reach the circumstantial conclusion that the benthic copepods inhabiting the Canal in 1902, when the pump plankton of Herdman was collected (see above) belonged to an essentially similar association as found today. From the two studies by Lipkin (1972a, b) it appears that the main water plant populations have also been constant since the list of Lyle and Rendle (in Fox, 1926) and Lami (1932), not to speak of Beets (1953). It is worthwhile to notice that we have no data prior to 1924 concerning the water-plant population of the Suez Canal. It is evident that at the time of Keller and of Krukenberg there was no *Halophila stipulacea* in the Canal and the advance of the macrophytic algae (besides perhaps *Cladophora*) was only beginning.

W. Steinitz (1929) considered the possibility of some species having later disappeared from the Canal waters. Gohar (1954) even implies that in the very first years of the Canal conditions were more propitious for settling of benthic animals since "the deposition of mud had not gone so far as to impede the settlement of plants and animals on the bottom". Taking up this theme Thorson (1971) considers that the "bottom which was fairly clean ... in 1869 (!) now consists of an increasing layer of gelatinous mud".

Probably most of the presumed cases of species' disappearance can be attributed to two circumstances: (1) unequal collecting (different sites, different methods, different seasons); (2) reduction of numbers owing to successional phenomena. W. Steinitz for instance gives the case of *Brachidontes variabilis* which was frequent in the Canal, but in the collections of the Cambridge Expedition it had become rare. He thought the same had happened to *Mactra olorina*. However, both molluscs are still very frequent in the Canal system. How little one can rely on the fortuitous lack or sparsity of a species in a collection can be shown by two examples: Gruvel (1936) found numerous populations of *Branchiostoma lanceolatum* in the Canal: the Cambridge Expedition, less than ten years before did not report one specimen. The same author reports the tectibranch gastropod *Notarchus savignyi* as being characteristically associated with *Halophila*, everywhere in Lake Timsah. This species does not appear in O'Donoghue's list (1929) based on the Cambridge collections, and Barash and Danin (1972) even believe that their record (under the name of *Bursatella leachi savigniana*) is the first report from the Suez Canal.

Only one instance given by W. Steinitz seems to be a case of a possible retreat or even disappearance: the rock-boring *Pholas dactylus* reported as very frequent by Keller and subsequently by Tillier and Bavay was not found by the Fox Expedition in 1924 "although a special search was made in the Canal banks". The sandstone slabs were found "riddled with holes presumably made by a boring mollusc but neither living molluscs nor shells were found within" (Fox, 1926, p. 29). Fox suggests "oil from ships" as one possible explanation for this disappearance. That especially in the artificial canal sectors, pollution and the Canal widening project could exterminate whole populations of sessile animals seems a reasonable assumption.

Succession also changed the quantitative appearance of some species. For instance, the intertidal levels were at an early stage inhabited only by *Balanus amphitrite* and *Brachidontes variabilis*. At the present, several other species have

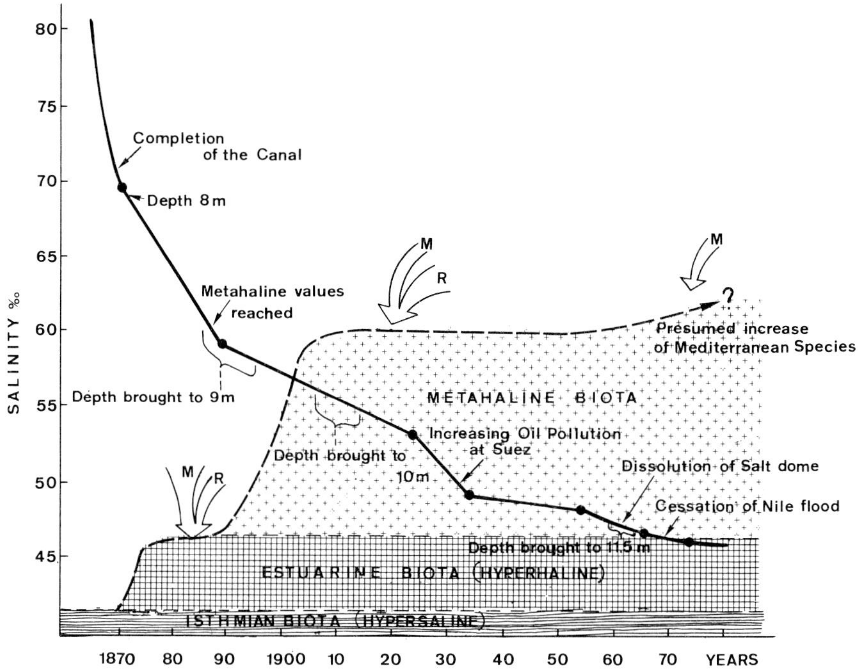

Fig. 28. A schematic representation of decreasing salinity and dredging events in the Suez Canal, versus gradual increase in species diversity. (original) (*M*: Mediterranean; *R*: Red Sea)

successfully carved out their living place in the same biotope; such later arrivals are the limpet *Cellana rota* and the barnacle *Chthamalus*. Consequently at least *Balanus amphitrite* has become rarer. Other species also, very frequent at the first stages, have become less obvious and have perhaps even retreated into the lagoons of Timsah or the Bitter Lakes. A good example of that is *Mactra olorina*.

Finally I would like to mention that Ben-Eliahu (1972d) discovered that between 1924 and 1967–69 the Mediterranean Polychaeta Errantia started to enter the Canal: in 1924 there was not a single species, whereas in the new collection seven species were found. Ben-Eliahu even suggests that the infiltration of the Mediterranean species into the Canal will increase in the future.

If this assumption is confirmed we shall have a fourth stage in the history of the Canal fauna, that of an increase of the Mediterranean influence. As the matter stands now, I do not see much reason to accept such a trend.

In conclusion, the Suez Canal started with a low-diversity autochthonous population of hypersaline–isthmus element–species. The first immigrants were hyperhaline–estuarine species. The second wave of immigrants were the metahaline marine species. At this stage the Canal biota reached a climactic stability corresponding to the near-stability reached by its salinity conditions. The main contingent of metahaline animals came from the Red Sea where such species have preexisted. An increase of the immigration of Mediterranean species will be eventually related to a more gradual advance and an evolutionary adaptation to high salinity metahaline conditions. This successional history is graphically represented in Figure 28.

Is the colonization of the Canal from the Gulf of Suez still going on? There are no data to indicate it. In fact the preadaptation to high salinities and the permanent tidal current reaching the Little Bitter Lake probably gave the chance for all potential invaders to reach the Canal. On the other hand, oil pollution has badly damaged the environment off Suez and also in the whole Gulf.

Gruvel (1936) more than forty years ago found that "... fish as well as invertebrates are repelled by the presence of oil from all the area near the Port Taufiq refineries and consequently from the entrance of the Canal". Gruvel states that the diversity of fishes has decreased in the Bay of Suez. As for the southern section of the Canal he writes "the number of fish and molluscs has diminished in this part of the Canal and in the lake from year to year". The decrease in the number of pearl oysters in the southern Canal is also due to the crude oil and refinery products, carried into the Canal by the tidal current: "they chase away the animals which can escape and destroy the sedentaries".

In the last decades the oil fields of El Morgan and Abu Rudeis have been increasingly operated on the Sinai coast of the Gulf. Most of these fields are submarine and big spills around the platforms are frequent. I had the occasion in 1974 to witness the heavy oil pollution caused by an accident on one of these platforms. More oil fields are operated on the African side of the Gulf.

It is, therefore, hard to conceive that new, i.e. more sensitive and stenotopic species, slower migrators, should have had better opportunities to colonize the Canal in the last years. Quite on the contrary, it is conceivable that some species which succeeded in gaining a foot-hold in the Southern Canal but could not reach the open expanses of the Bitter Lakes, were wiped out by the oil-laden tidal currents flushing past the refineries and oil terminals at Suez.

2.7 Species Diversity in the Suez Canal

This chapter considers the climax conditions presumably reached by the Suez Canal biota. Thorson (1971) calculated that the collections of the 1924 Cambridge Expedition contain a fauna of 318 species in the Canal. Lipkin (1972a) gives a list of 96 species of algae and sea grasses [if one disregards further three dubious findings by Muschler (1908)]. Por and Marcus (1972) give a total of 83 species of Copepoda Harpacticoida collected till now in the Suez Canal. Schellenberg's (1928) and Ruffo's (personal communication) lists brings the number of Amphipoda to 46 species. From different sources one may estimate that there are some 150 species of molluscs in the Canal. The report by Eitan (1972) brings the known species of the Bryozoa from the Canal, to 35 species. Ben-Eliahu (1972d) enumerates 78 species of Polychaeta Errantia. In Tillier's time (1902) there were 78 species of fish in the Canal; to this even the small collection reported by Steinitz and Ben-Tuvia (1972) adds another seven species.

One can thus assume that the total number of species in the Canal is in the magnitude of 1000–2000 species. Therefore, there is no reason to compare the

Suez Canal with typical and impoverished hyperhaline lagoons such as Laguna Madre or Sivash (Hedgpeth, 1957).

The Canal is a marine–metahaline environment. The species are almost exclusively of marine origin, however, the diversity is reduced because of a combination of high salinity–low temperature, adverse bottom and turbidity conditions. This is a typical metahaline situation. There are several animal groups in which the hyperhaline–estuarine species are also fairly well represented: fish, decapods, and copepods, as well as nematodes (Riemann and Rachor, 1972). Some taxa, such as Nemertes, Acarina, and Tanaidacea, etc. have still not been investigated. Because of the shallowness of the Canal system, animals living on deep bottoms are missing from the Canal fauna. The dearth of the plankton is however, a phenomenon much less suspected at first sight. Some typically holoplanktonic groups are missing from the Canal waters and the Canal lakes, for example Radiolaria, Foraminifera, Siphonophora, Ctenophora, planktonic Polychaeta, Myodocopid Ostracoda, Hyperiidea, Euphausiacea, Pteropoda, Heteropoda, Appendicularia, and Thaliacea.

Among the planktonic larvae no Echinodermata have ever been found in the Canal waters.

Holoplanktonic groups with a very restricted number of representatives in the Canal are Dinoflagellata, Tintinnida, Chaetognatha, Copepoda, and Cladocera.

Among the Dinoflagellata, Halim (1963), Kimor (1972), and Dowidar (1973) report *Ceratium furca*, *Ceratium aegyptiacum* and *Ceratium eupulchellum*. There are two tintinnids, *Tintinnopsis beroidea* and *T.radix* (Kimor, 1972). Burfield (1927) lists three arrow worms: *Sagitta enflata*, *S.neglecta*, and the neritic *Spadella cephaloptera;* there are no newer reports. Gurney (1927c) lists ten species of Calanoida which appear in the Canal waters, another five species were limited to the Canal entrances. From the Cyclopoida only *Oithona nana* spread in the Canal and from the Harpacticoida *Microsetella norvegica* and *Euterpina acutifrons*. Corycaeidae were found only in a few scattered specimens and no Sapphirinidae and *Copilia* were found.

Gurney (1927c) reports the Cladocera *Penilia avirostris* and *Evadne tergestina;* Kimor (1972) adds the remarkable *Bosmina coregoni maritima*, an estuarine cladoceran of freshwater origins. Among the Decapoda the planktonic *Lucifer hanseni* is found in the Lakes. In samples from December 1973 from the Great Bitter Lake, I found many specimens of Praniza larvae of gnathiid isopods and also Microniscus larvae.

The larval plankton is dominated by the swarms of nauplii of *Balanus*. Gurney (1927a) found larvae of Penaeidae, Alpheidae of Palaemonidae, of *Diogenes pugilator* and of *Leucosia signata*, all species also reaching maturity in the Canal system.

Schmidt (1972) considers that only the Hydrozoa with meroplanktonic larvae live in the Suez Canal and that therefore, they spread by means of short-lived medusae through the Canal.

Among the "epibenthic" planktonic organisms the Mysidacea are fairly common (Tattersal, 1927; Băcescu, 1973).

Kimor (1972) considers that the poor plankton of the Suez Canal is composed of euryhaline species, similar to those in estuarine environments, such as the

Sirbonic Lagoon (Bardawil), or the Qishon estuary near Haifa. He considers among the unfavorable conditions for the plankton life in the Canal, besides the high salinity and turbidity, "the shallowness of the Canal which almost excludes the passage, or even the temporary habitation of pelagic organisms committed to a certain degree of vertical migration". In my opinion vertical movements of the planktonic organisms are further hindered by the saline stratification of the waters in the Canal lakes.

It is interesting in this connection that the snail *Murex anguliferus*, which hatches already in the "swimming-crawling" stage and has probably a very short pelagic life, is widespread in the Canal. Another Red Sea species, *Murex ramosus* has a veliger with three days of pelagic life: this species has not penetrated into the Canal [letter by Thorson, Israel J. Zool. **21** (1972), p. 375]. Probably also the wide array of crab species which live in the littoral of the Red Sea often at fairly high salinities did not settle in the Canal because of their planktonic larval life: only one species of each Majidae, Grapsidae and Ocypodidae are widespread (Holthuis, 1956). After having listed 12 species of Echinodermata from the Canal (the crinoid *Lamprometra palmata* has been found only at Km 146), Mortensen (1926) discusses the fact that the widest-spread species in the Canal is the brittle star *Amphipholis squamata* which has no pelagic larvae. He writes: "pelagic larvae are not a necessary condition for rapid spreading (in the Canal)".

The combination of high salinity and turbidity is to blame for the peculiar composition of the fish-fauna and of the benthic fauna in the Suez Canal.

The *fishes* are first of all characterized by the euryhaline–estuarine species. These are still the numerically dominant group in the Canal. Despite the fear expressed in the Mediterranean world that the opening of the Suez Canal would result in an invasion by the man-eating Red Sea sharks, nothing similar happened. The Selachia are among the less-well represented animal groups in the Canal. There is in fact only one species of bottom sharks, *Himantura uarnak* which lives in the Canal. Gruvel (1936) speaks of occasional captures of hammer-sharks in the Bitter Lakes, but the report is not scientifically documented.

Understandably also high-sea pelagic fishes—such as Scombridae—are poorly represented in the Canal fauna. Exceptions among the pelagic fishes are *Sardinella aurita*, which penetrates from the Mediterranean, the euryhaline anchovy *Engraulis enchrassicolus*, *Pomatomus saltatrix* and the migrant *Dussumieria acuta*. The second and fourth of these species were reported in the Canal waters already by Tillier (1902).

Several families of fishes such as Labridae, Scaridae, Chaetodontidae, and Scorpaenidae, are not represented in the Canal fauna. One is inclined to think that the reason is first of all the lack of rocky bottoms and the associations of animals it carries, and especially the scleractinian reefs.

There are no *coral reefs* in the Suez Canal. From all the wealth of skeleton-forming Hydrozoa and Anthozoa, only the small gorgonarian *Acabaria erythraea* is found. This is an inhabitant of infralittoral boulders in the Red Sea, exposed to fluctuating environmental conditions. Dean (1929) reports that *Acabaria* was found growing on a sunken dredger which had been brought to Port Said, and eventually survived there for two months.

There are no Octocorallia in the Canal and from the rich fauna of Actinaria from the adjacent seas only four species settled in the canal by 1924 (Carlgren, 1927). One of these is the swimming sea-anemone *Boloceroides hermaphroditica* from the Red Sea, a highly mobile species.

The whole range of species connected with the coral reefs is consequently lacking from the Suez Canal. Barash and Danin (1972) give a whole list of missing families of molluscs: Architectonidae, Cypraeidae, Cymatiidae, Bursidae, Coralliophilidae, Mitridae, Conidae, Terebridae, and Tridacnidae. There are very few species of Polyplacophora, probably only one Scaphopod and not more than three species of Cephalopoda reported in the Canal. Here again the combination of salinity and edaphic conditions plays the principal role.

There are few species of Opisthobranchia in the Canal; however, some of them appear in great numbers, like the two species of *Berthellina*, *Bursatella leachi savigniana* (= *Notarchus savignyi*) and *Archidoris O'Donoghuei*. These are species living among the abundant growths of *Halophila* and the sponges which are associated with the creepers of this marine water plant.

Barash and Danin (1972) point out the interesting fact that in the Suez Canal there is a predominance of the number of species of Bivalvia over that of Gastropoda, despite the fact that in every sea the snails are usually more numerous than bivalves. This abnormal relation between the species diversity of the two main molluscan classes appears in all the collections made in the Canal since Tillier and Bavay (1905). The reason for this must be sought in the prevalence of the sedimentary bottoms over the hard bottoms.

At the other end of the animal kingdom, among the benthic Copepoda Harpacticoida, Por and Marcus (1972) remark that a whole array of rock-dwelling families are missing from the Canal waters: the Tegastidae, Peltidiidae, Porcellidiidae and to a great extent, the Tetragonicepsidae.

Generally speaking the Suez Canal environment is ideal for *shallow mixed-bottom species*. Among the molluscs, and more particularly gastropods, species living in the badly sorted soft bottoms of sandy-mud, such as *Murex anguliferus*, *Murex tribulus*, *Fusinus marmoratus*, *Cerithium scabridum*, and *Strombus tricornis* are extremely numerous in the Canal. In Gruvel's (1936) very detailed description of the soft-bottom associations of Lake Timsah, there are several species of lamellibranchs which appear in "quantités industrielles": *Gafrarium pectinatum*, *Callista florida*, *Tapes decussatus*, and *Mactra olorina*. Among the crabs, the Leucosiidae, a family typical of mixed bottoms are well represented by three species.

In the mixed and muddy bottoms of the Lakes several echinoderms are frequent, Synaptidae among the sea-cucumbers (3–4 species), *Astropecten polyacanthus* and several species of small brittle star (see also Beets, 1953).

Robinson (1927b) reports the Enteropneust *Dolichoglossus gurneyi* as inhabiting the shallow muddy bottoms of Lake Timsah sometimes in "enormous numbers". As already reported above, *Branchiostoma lanceolatum* is extremely abundant on the different types of sediment found in the Canal sectors, but not in the Lakes. Among the smaller benthos, Nematoda, Polychaeta, and Cumacea are very common, being typical for mixed bottoms. Among the harpacticoid copepods the level bottom species of the Cannuellidae, of the genus *Stenhelia*, as well as the very eurytopic mud-loving *Typhlamphiascus confusus* are very frequent.

Nearer to the shore a sequence of plant communities typical of other sheltered areas of the Red Sea is found. From the shore downwards these are *Digenea simplex*, *Halophila stipulacea*, *Halodule uninervis*, *Cystoseira*, and *Sargassum* spp. The littoral fauna associated with these plants is also the fauna found everywhere with them in the Red Sea. This so-called "*Halophila* fauna" and the neighboring "littoral boulder fauna" has been discussed to some extent in a previous paper (Por, 1971 b) as being extremely characteristic of the Bitter Lakes and Lake Timsah. The similarity of the Canal lakes' biota, with those of the semienclosed lagoons of the Northern Red Sea will be discussed below.

An impoverished fauna is found at the water level and in the belt corresponding to *the intertidal*, in the almost tideless canal water. Only those species of the Red Sea and Mediterranean intertidal are found which can survive or even thrive in sheltered environments. For instance, there are no representatives of the genus *Littorina* (sensu latu), no *Monodonta*, nor the big barnacles of the genus *Tetraclita*. Dominant there are *Brachidontes variabilis*, *Diodora rüppelli*, and *Balanus amphitrite*. Other species living in this belt are *Crassostrea (cucculata?)*, *Cellana rota*, *Modiolus auriculatus*, and *Chthamalus (stellatus?)*.

Finally, an important constituent of the Canal fauna is formed by the so-called "phytal fauna". Especially frequent are animals living on algae and water plants, and covering artificial substrates such as piers and buoys. These organisms participate in the fouling community found on ship hulls. Sponges, Hydrozoa, and Ascidiacea belong to this category. As for the smaller fauna, Pycnogonida, Hydracarina, and Amphipoda are widespread. Among the harpacticoid Copepoda the dominant ecological group is that of the phytal species (Por and Marcus, 1972). Among the Ostracoda also, the phytal species *Loxoconcha gardaqensis* is the most frequent (Lerner-Seggev, 1972). Of the ten species of Isopoda reported from the Canal waters (Glynn, 1972) seven or eight are typical of phytal habitats. Schellenberg (1928) in his exhaustive paper on the Suez Canal Amphipoda remarks the almost complete absence of the forms which burrow in the substrate. On the other hand, species "which climb among water plants and fixed animals" are well represented.

The Great Bitter Lake has sometimes been seen as an extreme hypersaline environment, with the connotation of low species diversity. Oren (1969) considered the Bitter Lakes as they were prior to 1924 "a solid barrier to the passage of organisms". To Thorson (1971) the Great Bitter Lake through a combination of hypersalinity with extreme high temperature (sic!) was an environment "problematic for adult animals and, probably, impossible for larvae". For a European marine biologist the salinity values of the Great Bitter Lake were hypersaline, and therefore beyond the range of the marine fauna. For the salinity-adapted Red Sea fauna this is different.

Keller (1882) found the fauna of the Bitter Lakes to be richer than that of the Bay of Suez. Fox (1929) concluded that the high salinity is not unfavorable to certain forms of marine life, and that the fauna of the Lakes is rich both in species and numbers of individuals. The teeming life in the Bitter Lakes is still a matter of surprise to everybody who visits its shores. In species diversity the Bitter Lakes can be compared well to a shallow and mixed-bottom gulf of a tropical sea.

2.8 Reproduction and Sizes of Animals in the Suez Canal

This chapter would have been largely unnecessary had it been clear to everybody that the Suez Canal is an environment to life as every other water body and not only a passage way, as it has been abstractly considered.

Time and time again, authors were surprised to report that they found reproductive or egg-bearing species in the Canal, on the assumption that the Canal system operates like a hypersaline or brackish lagoon, with adult populations either leaving for reproduction or having to be constantly reinforced by new swept-in swarms of larvae. Such a behavior can be expected from good swimmers: fish, penaeid shrimps, portunid crabs, or cephalopods. However, there are no indications that a migratory "catadromous" life cycle actually exists in any of these species in the Canal. The system, especially owing to the brackish areas of Lake Timsah, may in fact offer a very wide range of salinities for those species that need a certain salinity value in order to reproduce. Catadromous reproductive behavior is probably present but far from being the rule.

Among fishes, Tillier (1902) reports the following species as reproducing in the Canal: *Atherina forskalii (= Atherina pinguis), Clupea quadrimaculata, Crenidens forskalii (= C. crenidens), Mugil seheli, M. capito?* and *M. saliens?*. No additional nektonic fish in spawning stage has been reported since!

The nektonic decapods positively known to reproduce in the Canal are: *Penaeopsis stebbingi*, the larvae of which Gurney (1927a) found in the Canal, and the swimming crab *Thalamita poissoni*, which Calman (1927a) found to be ovigerous.

From the more "sedentary" organisms of the Canal relatively few have been reported to reproduce there. In fact, the collections are not specially analyzed for that; reporting of "adult" specimens is not enough. Fishes, for example, have to be dissected. *Hippocampus hippocampus* males were found with brood pouches full of developing eggs (Steinitz and Ben-Tuvia, 1972).

In molluscs the situation is similar for the bivalves; the egg-masses of Gastropoda can, however, be identified, especially by the large-sized spawns. The only cases in which reproduction of molluscs within the Canal could be proven are those reported by Barash and Danin (1972) in which egg masses were found, i.e. of *Murex anguliferus*, another species of *Murex* and an egg collar of *Natica* sp.

The situation is somewhat better in the case of benthic Decapoda; not only because most of them are egg-bearing animals, but also due to the fact that Gurney, the undisputed authority on decapod larvae, had had the opportunity to collect them personally. Larvae or egg-bearing females of 11 species of Decapoda were identified in the Canal waters (Balss, 1927; Calman, 1927a; Gurney, 1927a; Holthuis, in Por and Ferber, 1972). Gurney adds to the list another two unidentified types of crab larvae. If one considers, according to Holthuis (1956), that there are 38 species of Decapoda known from the Canal proper, then the 11–13 species known to reproduce there represent quite a high percentage. This is even more impressive if one takes into account that the Cambridge Expedition collections covered only the months of September and October. In the case of the ascidians of the Cambridge Expedition the ratio of the ovigerous species is equally high, i.e. eight out of 25 (Harant, 1927). The case of the Decapoda and Ascidiacea may indicate that reproduction in the Canal should be considered to be the rule rather than the exception.

Schellenberg (1928), analyzing a very diversified collection of Amphipoda, wrote that the populations in the Canal do not differ in fertility and in size from the populations outside the waterway. The same is true for the harpacticoid copepods, analyzed by Por and Marcus (1972).

The size problem has been mentioned in passing by Fox (1929). He set out to inquire whether the high salinity has a stunting effect on the Canal animals. The results of Fox indicate that the specimens living in the Canal are not smaller than their conspecifics living in the open sea. In some cases, such as the sea cucumber *Synaptula reciproquans*, the ascidian *Phallusia nigra* and the grey mullet *Mugil cephalus*, sizes are even greater than usual. No author ever reported having found specimens smaller than usual, or individuals deformed or deleteriously developed in any sense. One can, therefore, assume that the Canal populations live within their eco-physiological range. The populations living in the Canal reproduce normally, and are not dependent on occasional reinforcements coming from outside. The species living in the Canal are preadapted to the conditions which they encountered and there was no need for an adaptative process, or for appearance of "physiological races", nor of new genetic combination or subspecies. The Red Sea species inhabiting the Canal have behind them a long history of adaptation to high salinities in many different sites of their zoogeographical range.

2.9 The Metahaline Environments of the Red Sea and the Persian Gulf

To understand the success of the Red Sea species in the settling of the high salinity waters of the Suez Canal, one must for a moment raise the curtain over a much larger scene, that of the Western Indian Ocean. The hot desert shores of Arabia, the horn of Africa, and Persia shelter a wealth of semi-isolated littoral basins, with higher-than-sea salinities. If these basins are deep and permanently water-filled, they are termed "Ghor" or "Sharm". If they are shallow and periodically or permanently reduced to salt swamps, they are called "Sabkha". Evaporation may reach such high values as 300 cm/year.

In all these environments marine biota live, adapted unilaterally to high salinities. It is in a sense a hyper-eurhyalinity as opposed to the much better-known amphi-euryhalinity of the estuarine organisms.

The term *metahaline*, as proposed by Por (1972), covers environments of high salinity which are still inhabited by marine biota, fulfilling all their life-cycle in the respective basin. The marine hyper-euryhaline organisms of this type are consequently called *metahaline* organisms. The upper limit of the "Metahalinicum" has to be sought over 70‰ when through $CaCO_3$ precipitation the physico-chemical properties of the water change. The metahaline organisms are replaced there by amphieuryhaline organisms, many of which necessarily reproduce outside the high salinity basin.

At times, especially at low eustatic levels, whole portions of the adjacent seas of the Indian Ocean—the Red Sea and Persian Gulf—could turn into high-salinity metahaline bodies (Por, 1975a). The northern tip of the Red Sea, in the Gulf of Suez, reaches even today salinities of over 44‰. It is now generally accepted that during the Würm Glacial, the salinity of the Red Sea was even higher.

The shores of the Red Sea have many lagoon environments with salinities exceeding that of the open sea. Along the Sinai coast several such lagoons have been investigated more or less thoroughly: Ras Matarma; El Bilaiyim; the mangrove lagoons of Nabq; the pool of Di Zahav (Dahab) (see Map. 2).

The shallow Persian Gulf has salinities reaching 40‰ in the offshore waters. Along the southern coast, the Trucial coast, the salinity of the shallow shores reaches 50‰. In some lagoons values of even 70‰ are reached.

Along with the evaporation-induced high salinity, considerable fluctuation of temperatures in these metahaline environments also occurs. In some lagoons of the northern Red Sea, winter temperatures of 10° C have been measured. Even in the Persian Gulf sabkhas, winter temperatures of 15° C alternate with 40° C in the summer. The metahaline species are, therefore, adapted to minimum temperatures which are very much below tropical values. The adaptation to high salinity and low temperature can be considered together as a preadaptation to successful migration into the Suez Canal. Some of the above-mentioned environments and their biota, especially species also known in the Suez Canal, will be briefly presented below.

The Gulf of Suez. There are no comprehensive studies on the environments of this Gulf. With salinity increasing from 41‰ at the entrance of the gulf (maximum 60 m deep), to 44‰ at the northern tip, and the decrease of minimum winter temperatures from south to north, reaching a low 15–16° C, there is a gradual depletion of the tropical fauna. Coral reefs are found only in the southern sector of the Sinai coast. In the northern two-thirds of the coast, corals are adventive, young colonies i.e. coral communities instead of coral reefs (in the sense of Wainwright, 1965). The species diversity of the corals also decreases. This depletion is probably due not only to high salinity and low winter temperatures, but also to recurrent and catastrophic low tides which expose and kill the corals (Bannwarth, 1913) and to high turbidity. It goes without saying that many of the animals dependent on a flourishing coral reef do not reach the northern end of the Suez Canal.

From the many species of rock-living sea urchins of the Red Sea only two species, *Diadema setosum* and *Echinometra mathaei* reached the northern end of the Gulf of Suez according to James and Pearse (1969) Lawrence (1973) reported that these two species are more resistant to low temperatures than *Echinothrix calamaris* and *Tripneustes gratilla*, species which do not reach the Gulf of Suez. Even the widespread and common brittle star of the Red Sea, *Ophiocoma scolopendrina*, does not reach the Gulf of Suez (James and Pearse, 1969). The molluscs of the Bay of Suez, near the opening of the Suez Canal, have been well investigated since the times of Issel, Vaillant, Fischer, and Fuchs. The resulting data show clearly that the molluscs of the Canal represent an even more impoverished version of the Bay of Suez assemblage.

It is unfortunate for our subject that so little is known about the northern Gulf of Suez. However, it is evident that its environment is a first hurdle which the would-be settlers of the Canal have to pass on their way, a first sieve in the "sieving channel" of the Suez Canal.

Ghor Blayim also known as El Bilaiyim is a twin lagoon on the Sinai coast of the Gulf of Suez. The northern lagoon, 2 km long and 800–1000 m wide, has a depth of 15 m; the southern lagoon is smaller and shallower (up to 8 m). There is a 1–1.5 m deep narrow opening to the sea. The salinity reaches 48‰ in high summer while in the winter it is nearer to open sea salinities, i.e. 41–42‰.

The northern lagoon is more sandy, with *Tellina* sp. as dominant mollusc. In the southern lagoon, bottoms are predominantly rocky at shallow depths. Vegetation shows a typical zonation with *Digenea simplex* in the shallows, *Halophila stipulacea* and *Caulerpa racemosa* below, and *Sargassum* and *Halimeda* at depths of 2–3 m. There is a monospecific coral reef in the lagoon, formed of *Stylophora*, the most metahaline resistant Red Sea scleractinian. According to Loya (1976) this is a typical *r* strategist, a pioneer species. The soft corals are represented by only one species, *Sinularia*. Remarkable for our purpose is the fauna of molluscs and of fish. Besides *Spondylus* and *Chama broderipi*, which are not represented in the Suez Canal, the dominant molluscs are *Pinctada radiata*, *Mactra olorina*, *Malleus regula*, *Fusinus marmoratus*, *Strombus tricornis*, *Murex tribulus*, and *Diodora rüppellii*—all successful migrants into the Canal and even into the Mediterranean. The fish fauna, although not completely investigated, is characterized by estuarine species: outstanding is the presence of *Serranus cabrilla*, *Dicentrarchus punctatus*, and *Liza aurata*, three euryhaline species, two of them of Mediterranean origins (Ben-Tuvia, 1971b, 1975b, personal communication). According to Lipkin (1972b) the vegetation of the Bitter Lakes is comparable to that of the Bilaiyim lagoon, and represents a more hypersaline variety of the latter. The parasitofauna of the Bilaiyim fishes has much in common to that of the Bardawil lagoon (I. Paperna, personal communication).

Further investigation on the fauna of El Bilaiyim will no doubt increase the number of similarities with the Bitter Lakes. Por (1969a) for example found considerable similarities in the Canuellidae (Copepoda, Harpacticoida) inhabiting the two metahaline water bodies. Two of the isopods living in the Suez Canal, *Synidotea hirtipes* and *Cymodoce truncata* have also been recorded by Glynn (1972) from El Bilaiyim.

The Mangroves of Southern Sinai present salinities as high as 47‰ (Por and Dor, 1975b) and can also be considered as environments in which adaptation of Red Sea biota to metahaline conditions took place. Here again *Digenea* and *Halophila* are dominant plants, associated with other Suez Canal species: *Halodule uninervis*, *Spyridia filamentosa*, *Laurencia papillosa*, and *Caulerpa racemosa*.

Among the molluscs widespread in the Sinai mangrove we can mention again *Strombus tricornis* and *Fusinus marmoratus*, together with *Cerithium scabridum* and *Brachidontes variabilis*. The medusa *Cassiopea andromeda* and *Palaemon pacificus* are very common in between the mangrove trees and aerial roots.

Di Zahav pool (Por and Dor, 1975a) is a small (110 m × 70 m) pool on the Gulf of Elat coast of Sinai, which communicates with the open sea only through

Map 2. Israel and Sinai coasts (Israel: Cease Fire Lines, 1967) [from Por et al., 1972, reprinted from Israel J. Zool. **21** (3–4)]

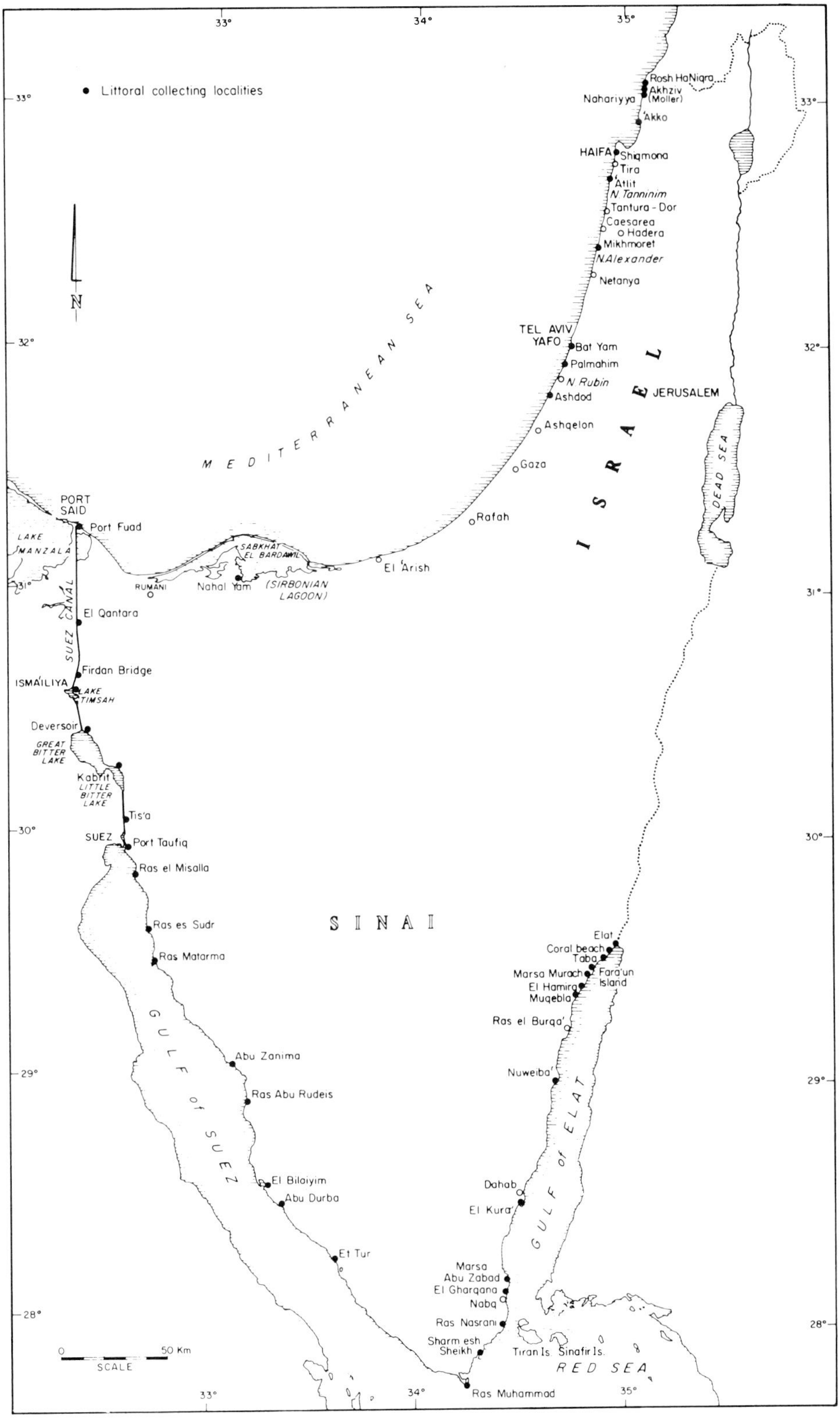

Littoral collecting localities
MEDITERRANEAN SEA
ISRAEL
SINAI
DEAD SEA
JERUSALEM
Rosh HaNiqra
Akhziv (Moller)
Nahariyya
'Akko
HAIFA
Shiqmona
Tira
'Atlit
N. Tanninim
Tantura - Dor
Caesarea
Hadera
Mikhmoret
N. Alexander
Netanya
TEL AVIV YAFO
Bat Yam
Palmahim
N. Rubin
Ashdod
Ashqelon
Gaza
Rafah
El 'Arish
SABKHAT EL BARDAWIL
(SIRBONIAN LAGOON)
Nahal Yam
RUMANI
PORT SAID
Port Fuad
LAKE MANZALA
SUEZ CANAL
El Qantara
Firdan Bridge
ISMA'ILIYA
LAKE TIMSAH
Deversoir
GREAT BITTER LAKE
Kabrit
LITTLE BITTER LAKE
Tis'a
SUEZ
Port Taufiq
Ras el Misalla
Ras es Sudr
Ras Matarma
GULF of SUEZ
Abu Zanima
Ras Abu Rudeis
El Bilaiyim
Abu Durba
Et Tur
Elat
Coral beach
Taba
Marsa Murach
Fara'un Island
El Hamira
Muqebla
Ras el Burqa'
Nuweiba'
GULF of ELAT
Dahab
El Kura'
Marsa Abu Zabad
El Gharqana
Nabq
Ras Nasrani
Sharm esh Sheikh
Tiran Is.
Sinafir Is.
RED SEA
Ras Muhammad
0 50 Km
SCALE
N
33°
34°
35°
32°
31°
30°
29°
28°

permanent seepage. Salinity fluctuates between 45‰ and 60‰ and winter temperature may be as low as 10° C. Nevertheless, among the relatively few species which live in this peculiar and extreme environment we can still find several species common also in the Suez Canal: *Diodora rüppelli* and *Ancillaria cinnamomea*, a small cowrie-like snail, as well as the medusa *Cassiopea andromeda*, the isopod *Cymodoce truncata* and the red alga *Laurencia papillosa*.

All the coastal shallows of the Red Sea may be considered as a metahaline area, with salinities reaching 45–46‰. This was evidenced for the first time by Remane and Schulz (1964) who briefly discussed the growths of *Digenea simplex* at Al Ghardaqa.

The Persian Gulf has extensive areas of high salinities along its southern shores. The biota of the Gulf shows a gradual depletion with increasing salinity, a "restriction" as termed by Purser and Seibold (1973). The restriction in the fauna is shown in the open waters of the Gulf where such Indo-Pacific genera as *Tridacna* or *Lambis* do not occur. According to Hughes and Kay (1973), *Conus* and *Strombus*, all the pectinid shells and almost all of the sea urchins disappear at salinity values of 45‰. Of all the coral species only *Porites* advances to 48‰. However, marine biota can be still found at higher salinities. Among them are many of the Suez Canal species. According to Kessler (1973) the sea star *Asterina wega*, *Astropecten* (probably *irregularis*) and an unidentified brittle star can still be found at > 50‰ in Ghor al Bazam.

In the Abu Dhabi lagoon Evans et al. (1973) report aquatic plants such as *Halophila stipulacea*, *Halodule uninervis*, *Jania* sp. and *Acetabularia* sp. from salinities reaching 50‰. Several species of molluscs still live on the intertidal flats of the inner lagoon with salinities of 53.6–66.9‰, among them *Cerithium scabridum*, *Brachidontes variabilis* and *Pirenella conica*. It is interesting to note that the most salinity-resistant of the sea urchins on the Abu Dhabi shore is *Echinometra mathaei*, which lives at salinities of 42.7–44.5‰, a similar situation to that in the northern Gulf of Suez.

The coastal metahaline environments of the Indian Ocean and the adjacent seas, the Red Sea and the Persian Gulf, are still insufficiently investigated. It seems, however, reasonable to assume that more and more species known in the Suez Canal will be found also in these other high salinity environments. It is evident that large parts of the Canal fauna are in fact species which were "adaptatively ready" to invade the artificial metahaline environment built by de Lesseps.

3. The Migrant Biota

3.1 What is a Lessepsian Migrant?

The term "Lessepsian" was first used by Por (1964) to characterize the new phase into which the Eastern Mediterranean had entered with the opening of the Suez Canal. Por (1969a, 1971b) coined the term "Lessepsian migrant" for Red Sea species which have passed through the Suez Canal and settled in the Eastern Mediterranean.

In zoogeography the phenomena are usually studied post factum and the argumentation is circumstantial. The opening of the Suez Canal and the first reports on erythraean immigrants in the Mediterranean kindled the enthusiasm of zoologists. The Suez Canal became a link to justify the presence of tropical species even in the Black Sea or the finding of Atlantic species in the southern Red Sea.

A primary task for anyone who studies this subject is to weed out the considerable number of would-be migrants, according to a set of criteria as reliable as possible. This has been attempted on two previous occasions (Por, 1971b, 1973b).

I shall try to point out the different categories of reasoning I have used to reject would-be migrants, and to differentiate between presumable and certain Lessepsian migrants. It would be presumptuous to claim that the judgements proposed are of absolute value. However, every phenomenon can be studied only on the basis of a set of working principles.

Circumtropical Species. These are present ab initio on both sides of the Suez Canal. This is true in the case of many planktonic species: Kimor (1973) writes that the microplankton of the Mediterranean and the Red Sea "is characteristic of all warm-water seas in the world and therefore almost circumtropical in distribution". Even in cases like the Tintinnidae, in which according to Kimor (Komarovsky, 1959) the resemblance between the Red Sea fauna and the Eastern Mediterranean fauna is greater than between the Eastern and Western Mediterranean; Kimor (1973) attributes this to the effect of resembling abiotic factors rather that to a migrational effect. Alvariño (1974) assures that the presence of the siphonophorans *Chelophyes appendiculata* and *Sulculeolaria chuni* on both sides of the Suez Canal is due to cosmopolitan distribution and that "the source of the populations in the regions adjacent to the Suez Canal may be in the adjacent regions". Circumtropical, or tropicopolitan species among algae have frequently been considered as Lessepsian migrants, e.g. the species of *Caulerpa* found in the Eastern Mediterranean (Hamel, 1930, 1931; Rayss, 1941). Many of the weed-haunting harpacticoids (Por and Marcus, 1972) are circumtropical or even cosmopolitan. There are also some cases in which a circumtropical distribution has become evident only recently—either due to more thorough collecting, or to a taxonomic revision. Such may be the case of the little pearl oyster *Pinctada radiata* which is circumtropical according to its presence along the western shores of America and its wide distribution in the Western Mediterranean. The fishes *Carcharhinus brevipinna*, *Etrumeus teres*, and *Parexocoetus mento* which are often considered to be migrants from the Red Sea are, in fact, circumtropical species (Ben-Tuvia, 1971a).

Circumafrican Species. These are species found not only in the Mediterranean and the Red Sea, but also along the Western shores of Africa, and some of these species are, in fact, circumtropical. These species had ample opportunities to reach the Mediterranean through the Straits of Gibraltar or through previous connections of the Mediterranean with the Atlantic. *Aspidosiphon elegans*—a representative of a very old genus of Sipunculida, is considered by Wesenberg-Lund (1957b) to be a migrant through the Suez Canal. However, this species is also known from West African shores, from whence it probably reached the Mediterranean.

Species Not Found in the Vicinity of the Canal. Here the problem is more complicated. There are several instances in which a species formerly considered to be Atlanto–Mediterranean is also reported from the Southern Red Sea. Some of the species of Polychaeta, reported by Fishelson and Rullier (1969) as having migrated through the Suez Canal, have been reported from the Southern Red Sea. The bopyrid *Parathelges racovitzai* reported by Codreanu (1967) from the Black Sea as an Indo-Pacific element cannot be reasonably considered as a Lessepsian migrant since the distance it had to cover and the extremely different conditions to which it would have had to adapt obviously rule this out.

Schmidt (1972) reports five species of Atlanto–Mediterranean hydroids from Elat. He considers the possibility of their having used passive transport (on ship hulls) in order to circumnavigate the Sinai peninsula. This is, of course, a possibility which cannot be ruled out, although it seems much more probable that the species are and were present in the Red Sea before the opening of the Canal.

Pre-Lessepsian Species. Although information on species distribution before the opening of the Suez Canal is scarce, in some instances it serves to disqualify would-be Lessepsian migrants. Such is the case of the polychaete *Eurythoë complanata*, a species widespread in the Indo-Pacific which, however, was reported from Beirut before the opening of the Canal (Fauvel, 1955). The fish *Serranus cabrilla*, an Atlanto–Mediterranean species, usually considered to be a migrant through the Suez Canal, was reported from the Red Sea before the opening of the Canal (Klunzinger, 1884, based on Hemprich and Ehrenberg).

The euryhaline species, called "Isthmus species" (see above), such as *Cerastoderma glaucum*, *Pirenella conica*, *Aphanius dispar*, and eventually *Mugil auratus* and *Dicentrarchus* spp., could have crossed the isthmus before the shipping channel was opened.

Ship-Fouling Fauna. As remarked in a previous paper (Por, 1973b), some of the recently disseminated fouling organisms of eastern origin, such as *Mercierella enigmatica* and *Elminius modestus* were also probably carried through the Suez

Canal. The same may be true for some bryozoans. The wood-boring amphipod *Chelura terebrans* has to be considered among the passively transported fouling organisms which do not need the special opportunity provided by the Suez Canal in order to spread to the limits of their environmental tolerance.

Accidental Expatriates. The two species reported by Babič (1911) from the port of Fiume (Rijeka) and collected there in 1896 (the swimming crab *Neptunus sanguinolentus* and the rock lobster *Thenus orientalis*) evidently reached the port through accidental transport. The finding of one specimen of the Red Sea fish *Abudefduf saxatilis* in the Gulf of Naples (Tardent, 1959) may be included in the same category. In molluscs the possibility of the transportation of shells by man for cultural purposes has to be kept in mind. Haas (1948) found an empty shell of the cowrie *Monetaria moneta* near Acre and admits to the possibility that it served to decorate the harness of a horse or a camel.

Errors and Incomplete Identifications. The pencil urchin *Heterocentrotus mammillatus* has been reported by Gauthier (1874) and Carus (1893) as "being in the process of invading the Mediterranean". No tangible proof for this finding was ever produced and it has to be considered as due to erroneous locality indication. Pesta (1918) reports the finding of a specimen of the deepsea *Platymaia wywillethompsoni* from the Dalmatian port of Kotor in the Museum of Vienna. W. Steinitz (1929) suggests that there was a mistake in the museum label. According to Holthuis (1956) the report of *Porcellana boscii* in the Mediterranean is based on a misidentification. Pérès (1967) lists the cirripede *Tetraclita* sp. among the Lessepsian migrants, however, I could not find the reference for this finding, nor a final identification.

A last example is Haas (1942) who reported a benthic comb jelly *Coeloplana* sp. from the Mediterranean coast of Israel. The genus is still considered an exclusively Indo-Pacific genus. However this unidentified ctenophore of the Mediterranean should not be considered as a Lessepsian migrant only on this ground.

In all the above-mentioned categories, true Lessepsian migrants might eventually be found. This is especially the case for the circumtropical and circumafrican species. Some of these might actually be absent from the Eastern Mediterranean or the Northern Red Sea and might have used the Suez Canal for their present distribution. In other cases, a circumtropical species might be found to have subspecifically segregated populations at both ends of the Canal; a more detailed study of such a species may prove in the future that one of the subspecies has actually invaded the area of another.

The restricted number of evident and demonstrable Lessepsian migrants is therefore the result of a very careful selection (see Table 2). It is reasonable to assume that some of the more dubious cases—listed separately (Table 3)—are also true Lessepsian migrants. Finally one has to take into account that a thorough investigation of areas and taxa insufficiently studied will probably treble the number of actually known cases.

In the following list of evident migrants three categories have been included representing different degrees of a high probability of migration.

1. Species known from the Red Sea, and from the Suez Canal, and known to have spread along the Mediterranean coasts.

2. Species known from the Red Sea and the Canal, but still confined to "take off" areas off Port Said or in the Sirbonic lagoon.

3. Indo-Pacific species known from the Red Sea, reported also in the Eastern Mediterranean, but not found in the Suez Canal.

A special category is formed by the parasitic species, where for instance, the proof for the migration of the fish species, served also as proof for its parasite.

The question may rightly be posed as to why the above listing of proven Lessepsian migrants considers only a certain unidirectional movement from the Red Sea to the Mediterranean. There are only a few instances of proven migration in the opposite direction, i.e. from the Mediterranean to the Red Sea.

For the sake of clarity the name "Lessepsian migrants" refers to the bulk of species which migrated northward, while the very few species which moved south are called here "Anti-Lessepsian migrants". A special chapter will deal with the analysis of the overwhelming unidirectionality of the migration through the Canal.

H. Steinitz (1967) published a list of migrants called *A Tentative List.* In a research report of the Hebrew University—Smithsonian Institution Joint Program *Biota of the Red Sea and the Eastern Mediterranean* (1970), H. Steinitz compiled a revised list entitled *Comprehensive List of Immigrant Animals.* In the same report he started another type of list called *A Critical List of Immigrants via the Suez Canal* of which only the tables for Porifera, Polychaeta, and Echinodermata were completed. I have consulted some of the notes for the other taxa, left by Steinitz, for the list below.

The following list will deal with plants and animals, and presents separately the evident and doubtful cases of Lessepsian migrants.

A third list of accidental or highly improbable cases is listed in succinct form (Table 4).

Finally, an approach which does not seem to be evenhanded has to be justified: in well investigated taxa, such as fishes, molluscs, echinoderms, I have taken less care in including a species among the certain Lessepsian migrants. Yet the critical selection has been much more severe among the ascidians, decapods and algae, and most emphatically among polychaetes, bryozoans, sponges, and copepods.

The first two lists contain 128 species considered to be certain Lessepsian migrants and 76 species considered to be uncertain. Dowidar (1973, 1974) speaks of four species of phytoplankton and seven species of Tintinnida which have migrated into the Mediterranean. There are several species of migrant fish parasites under study (Paperna, personal communication). No less than five species of fish have been added to the list of migrants in the last few years. Even if this does not necessarily indicate an ongoing immigration it certainly points to the fact that even in the well-investigated taxa, many migrants are still undetected. Finally, one has to add that in many animal groups, such as Platyhelminthes, Nematoda, Acarina, and Hydrozoa, the Eastern Mediterranean is still a "terra incognita". The list of H. Steinitz (1967) contained 114 species of Lessepsian migrants, some ten of them uncritically included. One may thus say that in less than ten years, the list has doubled due in most part to new reports of migrants.

Table 2. High probability Lessepsian migrants (HPLM)

Category[a]	Taxon and species (species in alphabetical order)	Record of distribution and authority			Synonymy and Remarks
		Mediterranean	Suez Canal	Red Sea and Indo-Pacific	
	Algae				
	Diatomacea				
	Chaetoceros coarctatus	Haifa Bay (Kimor, pers. comm.)	— ? —	Red Sea (Kimor, pers. comm.)	In both localities with an epiphytic *Vorticella*
	Rhizosolenia indica	Israel coast, Bardawil Lagoon (Kimor, 1973)	Suez Canal (Ghazzawi, 1936)	Indo-West-Pacific	—
	Macrophyta				
3.3.1	*Acanthophora najadiformis* (Delile) Papenfuss	Alexandria (Aleem, 1948)	Suez Canal	Suez, Ghardaqa, Aqaba, Jiddah, Hodeida	= *A. Delilei* (Aleem, 1948: "The migration could have taken place in historical times through the old intermittent canal")
2.3.2	*Acetabularia calyculus* Quoy et Gaimard	Israel (Rayss, 1955) Bardawil Lagoon (Lipkin, 1972a)	Great Bitter Lake: 1967 (Lipkin, 1972a) ? (Lyle in Fox, 1926)	Suez, N. Red Sea (Nasr, 1947; Papenfuss, 1968)	= *A. mediterraneum*? = *A. acetabulum* of Lyle (Lipkin, 1972a)
4.2.4	*Acetabularia moebii* Solms-Laubach	Alexandria (Aleem, 1948) Algeria (Feldmann and Feldmann, 1947) Israel (Rayss, 1955)	— ? —	Ghardaqa (Nasr, 1947) → Indo-West-Pacific	= *A. Wettsteini* Naples (Schussnig, 1930)
4.2.4	*Cladophoropsis zollingeri* (Kütz.) Börgesen	Alexandria, Sallum (Aleem, 1948) → Rosh Haniqra (Rayss, 1955)	— ? —	Red Sea, Persian Gulf → Indo-West-Pacific	Cape Horn (? — unlikely)
3.3.2	*Hypnea esperi* Borg.	Israel (Lipkin, 1972a)	Little Bitter Lake 1967, 1969 (Lipkin, 1972a)	Gulf of Elat (Rayss and Dor, 1963)	Fertile plants in the Suez Canal (Lipkin, 1972a)
3.3.2	*Hypnea valentiae* (Turner) Montagne	Rhodes (Reinbold, 1898)	(Lyle in Fox, 1926)	Ghardaqa, Aqaba, Tor (Papenfuss, 1968)	"Possible transport by ships" (Fritsch, 1895)

[a] See key to symbols in Table 5.

Table 2 (continued)

Category[a]	Taxon and species (species in alphabetical order)	Record of distribution and authority			Synonymy and Remarks
		Mediterranean	Suez Canal	Red Sea and Indo-Pacific	
4.2.1	*Rhodymenia erythraea* Zanardini	Port Said Harbour (Aleem, 1948)	—?—	Yemen (Zanardini, 1858) Erithrea; Karachi	—
4.3.2	*Sarconema filiforme* (Sond.) Kylin	Palmahim → Caesarea (Rayss, 1963)	—?—	Yemen, Karachi, Bombay	"Tethysrelikt" (Rayss, 1963)
4.2.1	*Sarconema furcellatum* Zanardini	Port Said, Port Fuad (Aleem, 1948)	—?—	Red Sea, Arabian Sea, Ceylon	—
3.3.1	*Solieria dura* (Zanardini) Schmitz	Bat Yam 1957, 1960 (Rayss, 1963) Port Said (Aleem, 1950)	Ismailia, Port Taufiq (Aleem, 1950)	Yemen, Djibouti	= *Rhabdonia dura* Tethys relict (Not in the Red Sea [!?]: Rayss, 1963)
4.2.1	*Spatoglossum variabile* Figari et de Notaris	Port Said → El Arish (Aleem, 1950)	—?—	Suez, Aqaba, Ethiopia (Papenfuss, 1968)	—
	Cormophyta				
2.1.3	*Halophila stipulacea* (Forsskål) Ascherson	Rhodes 1894 (Fritsch, 1895) W. Egyptian coast (Aleem, 1962) Aegean Sea (Pérès, 1967)	1924 (Fox, 1926) Bitter Lake, Timsah (Beets, 1953; Lipkin, 1972a)	Gulf of Suez, Gulf of Elat, Indo-West-Pacific	Most authors: "Lessepsian migrant" (Pérès, 1967; Por, 1971b: "pre-Lessepsian")
	Animalia				
	Porifera				
4.2.2	*Chrotella cavernosa* (Lamarck)	Alexandria (Burton, 1936) Israel (HUJ collections) Israel coast (Tsurnamal, 1969a)	—?—	Red Sea, Indo-West-Pacific	= *Tethya* = *Cinachyra australiensis* "Dominant element in caverns and nips of Israel coast" (Tsurnamal, 1969a)
3.3.2	*Damiriana schmidti* (Ridley)	Israel: Mikhmoret (Tsurnamal, 1969a)	Suez Canal (Tsurnamal, personal communication)	Southern Red Sea (Levi, 1958, 1965) → Hawaii	= *Crella* = *Dendoricella* = *D. australiensis* = *D. hawaiana*

4.3.2	*Geodia micropunctata* Row	Israel: 'Akko (Tsurnamal, 1969a)	— ? —	Buoy in Port Suez (Row, 1911) Dongonab, Red Sea	—
3.3.2	*Heteronema erecta* Keller	Israel: Yavne, 'Akko (Tsurnamal, 1969a)	Great Bitter Lake, El Qantara (Tsurnamal, personal communication)	Red Sea (Levi, 1965) Indian Ocean	—
2.2.1	*Reniera spinosella* Row	Port Said (Burton, 1926)	Km 46, Km 152 (Burton, 1926)	Port Taufiq; Red Sea	—
	Annelida: Polychaeta				
2.3.2	*Branchiosyllis uncinigera* Hartmann-Schröder	Israel (Harlock and Laubier, 1966) Mikhmoret (Lipkin and Safriel, 1971)	Km 46 (Fauvel?, 1927) Great Bitter Lake, El Qantara — 1967 (Ben-Eliahu, 1972d)	Sarso Island	—
3.3.1	*Glycinde bonhourei* Gravier	El Arish (Ben-Eliahu, 1972c)	El Qantara — 1967 (Ben-Eliahu, 1972d)	Djibouti (Gravier, 1906a)	—
2.3.2	*Hydroides heteroceros* (Grube)	Beyrouth (Laubier, 1966) Israel? (Ben-Eliahu, 1972c)	Km 46, Km 152 (Potts, 1928)	Gulf of Suez (Fauvel, 1933) Persian Gulf, Zanzibar, India	= *H. uncinata* = *Eupomatus*
2.1.2	*Nereis persica* Fauvel	Port Said (Fauvel, 1927) Sinai (Ben-Eliahu, 1972d) Israel? (Laubier, 1966)	El Qantara (Fauvel, 1927) Little Bitter Lake, El Qantara (Ben-Eliahu, 1972d)	Port Taufiq, Persian Gulf → Indo-West-Pacific	= *N. zonata persica*
2.2.1	*Nereis willeyi* Day	Alexandria (Fauvel, 1937)	Suez Canal "frequent" (Fauvel, 1927) *N.* cf. *willeyi*: Great Bitter Lake (Ben-Eliahu, 1972d)	Persian Gulf, Indo-West-Pacific	= *N. capensis*
2.1.1	*Perinereis nuntia typica* (Savigny)	Port Said (Eastern jetty) (Fauvel, 1927)	Timsah, Bitter Lake (Fauvel, 1927) Great Bitter Lake (Ben-Eliahu, 1972d)	Port Taufiq (Fauvel, 1927; Ben-Eliahu, 1972d) Persian Gulf, Indian Ocean	—

[a] See kay to symbols in Table 5.

Table 2 (continued)

Category[a]	Taxon and species (species in alphabetical order)	Record of distribution and authority: Mediterranean	Record of distribution and authority: Suez Canal	Record of distribution and authority: Red Sea and Indo-Pacific	Synonymy and Remarks
2.1.2	*Pseudonereis anomala* Gravier	Alexandria (Fauvel, 1937) Israel (Fauvel, 1955) Lebanon (Laubier, 1966) Cyprus (Ben-Eliahu, 1972 b)	Km 146 (Fauvel, 1927)	Gulf of Elat (Ben-Eliahu, 1972 d) Red Sea (Gravier, 1906 a)	—
2.1.2	*Syllis exilis* Gravier	Port Said (Fauvel, 1927) Israel coast (Monro, 1937; Tebble, 1959)	Km 24 (Fauvel, 1927)	Red Sea (Gravier, 1908)	—
2.3.2	*Terebella ehrenbergi* Gravier	Cyprus (Ben-Eliahu, 1972 b)	Km 14 (Potts, 1928)	Gulf of Suez (Potts, 1928) Red Sea (Gravier, 1908)	Western Mediterranean (Pérès, 1954)
	Pycnogonida				
2.3.2	*Anoplodactylus saxatilis* Calman	Israel: Dor (Stock, 1958)	Little Bitter Lake (Calman, 1927 c)	Port of Suez, Indian Ocean, Singapore	= *A. digitatus*
	Crustacea: Copepoda				
2.3.2	*Acartia centrura* Giesbrecht	Israel (Berdugo, 1974)	→ Ras el Ech (Gurney, 1927 c)	Bay of Suez (Gurney, 1927 c) Red Sea, Indian Ocean	—
1.3.2	*Calanopia elliptica* (Dana)	Israel: near Haifa (Berdugo, 1968)	Suez Canal: also near Port Said (Thompson and Scott, 1903)	Indo-West-Pacific	—
3.3.2	*Calanopia media* Gurney	Israel (Berdugo, 1968)	Suez Canal (Gurney, 1927 c)	Indo-West-Pacific	—
2.3.1	*Canuellina insignis* Gurney	Sirbonis (Bardawil Lagoon) (Por, 1972)	Port Said (Gurney, 1927e) Little Bitter Lake → El Qantara (Por and Marcus, 1972)	Gulf of Suez, South Red Sea (Por, 1969 a) Inhaca	—

4.3.1	*Enhydrosoma vicinum* Por	Sirbonis (Bardawil Lagoon) (Por, 1972)	— ? —	Gulf of Elat (Por, 1969 a)	—
4.3.1	*Paramphiascella sirbonica* Por	Sirbonis (Bardawil Lagoon) (Por, 1972)	— ? —	Di Zahav Pool (Por and Dor, 1975 a)	—
2.3.1	*Robertsonia salsa* (Gurney)	Sirbonis (Bardawil Lagoon) (Por, 1972)	Suez Canal (Gurney, 1927d; Por and Marcus, 1972)	Solar Lake (Por, 1972) Di Zahav Pool (Por and Dor, 1975a)	—
3.3.2	*Scottolana longipes* (Thompson et A. Scott)	Israel Coast (Por, 1964) Bay of Pelusium (Por, 1969a)	Great Bitter Lake (Por and Marcus, 1972)	El Biliayim (Por, 1969 a)	= *Canuella* Possible pre-Lessepsian element (Por, 1969a)
3.3.1	*Stenhelia inopinata* A. Scott	Sirbonis (Bardawil Lagoon) (Por, 1972)	Little Bitter Lake → El Qantara (Por and Marcus, 1972)	Port Taufiq (Por and Marcus, 1972) Indian Ocean	—
2.3.1	*Stenhelia minuta* A. Scott	Israel coast (Por, 1964)	(Gurney, 1972e; Por and Marcus, 1972)	Elat (Por, 1967) Gulf of Suez Indian Ocean	Possible pre-Lessepsian (Por, 1964)
	Crustacea: Tanaidacea				
3.3.2	*Kalliapseudes omer-cooperi* Larwood	Gaza (Băcescu, 1961)	Port Taufiq (Canal outlet) (Larwood, 1954)	Gulf of Suez (Plankton) (Larwood, 1954) Indo-Pacific	—
	Crustacea: Stomatopoda				
2.2.2	*Squilla massavensis* Kossman	Alexandria (Steuer, 1938) Israel coast 1955 (Ingle, 1963)	Great Bitter Lake 1932; Timsah 1933; Km 4–5, 1934 (Dollfus, 1938; Tortonese, 1952)	Red Sea, Indian Ocean	—
	Crustacea: Decapoda				
2.2.2	*Alpheus audouini* Coutière	Israel 1951 (Gottlieb, 1953)	Kabret → Port Said (Balss, 1927; Gurney, 1927a; Tortonese, 1952)	Red Sea, Indian Ocean	Larvae in Suez Canal (Gurney, 1927a)
3.3.4	*Alpheus crassimanus* Heller	Alexandria (Balss, 1936) Tunisia: Sfax (Forest and Guinot, 1958)	(Gruvel, 1936; Monod, 1937)	Red Sea, Indo-Pacific	—

[a] See key to symbols in Table 5.

Table 2 (continued)

Category[a]	Taxon and species (species in alphabetical order)	Record of distribution and authority			Synonymy and Remarks
		Mediterranean	Suez Canal	Red Sea and Indo-Pacific	
4.3.2	*Atergatis roseus* (Rüppell)	Israel: Tel Aviv, Mikhmoret 1961; Israel to Lebanon (Lewinsohn and Holthuis, 1964)	—?—	Red Sea, Indo-Pacific	—
4.1.2	*Charybdis helleri* (A. Milne-Edwards)	Haifa Bay (Steinitz, 1929) Jaffa 1929 (Monod, 1930) Alexandria (Balss, 1936)	—?—	Red Sea, Indo-Pacific	= *C. (Goniosoma) merguiensis* Must have entered E. Mediterranean through Suez Canal (Holthuis and Gottlieb, 1958)
4.2.3	*Charybdis longicollis* Leene	Mersin Bay 1954 (Holthuis and Gottlieb, 1956) Israel 1961 (Lewinsohn and Holthuis, 1964)	—?—	Red Sea, Persian Gulf	May be a Tertiary relict (Tortonese, 1969)
2.2.1	*Eucrate crenata* (de Haan)	Port Said (Calman, 1927a) Alexandria (Balss, 1936)	(Calman, 1927a; Monod, 1938; Tortonese, 1947, 1952; Holthuis, 1956)	Red Sea → Indo-Pacific	—
2.2.1	*Heteropanope laevis* (Dana)	Port Said (Calman, 1927a) Egypt (Holthuis, 1956)	(Calman, 1927a) Little Bitter Lake (Gruvel, 1936) Timsah (Monod, 1938)	Red Sea, India	= *Pilumnopeus*
2.3.2	*Hyastenus hilgendorfi* de Man	Israel: Ashdod—Bat Yam 1960 (Lewinsohn and Holthuis, 1964)	(Calman, 1927a; Gruvel, 1936; Monod, 1938; Por and Ferber, 1972)	Red Sea, Indo-Pacific	—
4.2.1	*Leptochela aculeocaudata* Paulson	Alexandria (Balss, 1936)	—?—	Red Sea	—

2.3.2	*Leucosia signata* (Linnaeus)	Israel: Bat Yam 1953 (Holthuis and Gottlieb, 1958; Lewinsohn and Holthuis, 1964)	Ballah (Calman, 1927a) (Monod, 1938; Holthuis, 1956; Por and Ferber, 1972)	Red Sea → Zanzibar	Larvae in Suez Canal (Gurney, 1927a)
2.2.3	*Metapenaeus monoceros* (Fabricius)	Abukir (Balss, 1936) Israel 1949–1950 (Wirszubski, 1953) Iskenderun, Antalya (Geldiay and Kocatas, 1972)	→ Port Said (Balss, 1927; Gurney, 1927a) (Burkenroad, 1934; Gruvel, 1936)	Red Sea	= *Penaeopsis*
1.2.2	*Metapenaeus stebbingi* Nobili	Abukir (Balss, 1936) Haifa 1958 (Lewinsohn and Holthuis, 1964) Sirbonis (Bardawil Lagoon) (Por, 1972)	(Krukenberg, 1888a; Balss, 1927; Gurney, 1927a; Gruvel, 1936; Monod, 1937; Por and Ferber, 1972)	Red Sea, Mozambique	= *Penaeus;* = *Penaeopsis*
2.2.3	*Myra fugax* (Fabricius)	Jaffa (Monod, 1930) Iskenderun 1930, Alexandria 1936 (Balss, 1936)	→ Ballah (Calman, 1927a; Fox, 1926) Monod, 1937, 1938)	Red Sea, Indo-Pacific	Western Mediterranean? (Gilat, 1969)
4.3.2	*Notopus dorsipes* (Linnaeus)	'Atlit (1 specimen) 1962 (Lewinsohn and Holthuis, 1964)	— ? —	Gulf of 'Aqaba (Monod, 1938) Ghor Dongonab (Laurie, 1915) Indo-West-Pacific	—
4.2.2	*Palaemonella vestigialis* Kemp	Nahariya 1951, Haifa 1953 (Holthuis and Gottlieb, 1958)	— ? —	Gulf of Suez (Kemp, 1922) Red Sea, Indo-Pacific	"Most probable migrant through Suez Canal" (Holthuis and Gottlieb, 1958)
2.2.3	*Penaeus japonicus* Bate	Syria (Gruvel, 1928) Iskenderun (Monod, 1930; Gottlieb, 1953; Geldiay and Kocatas, 1972)	→ Port Said (Balss, 1927; Gurney, 1927 a; Gruvel, 1936)	Red Sea, Indo-Pacific	= *P. canaliculatus* Frequent in catches

[a] See key to symbols in Table 5.

Table 2 (continued)

Category[a]	Taxon and species (species in alphabetical order)	Record of distribution and authority: Mediterranean	Suez Canal	Red Sea and Indo-Pacific	Synonymy and Remarks
3.2.3	*Penaeus semisulcatus* de Haan	Syria (Gruvel, 1928) Iskenderun (Monod, 1930; Geldiay and Kocatas, 1972) Israel (Gottlieb, 1953; Wirszubski, 1953)	"Tout le Canal" (Gruvel, 1936; Monod, 1937)	Red Sea, Indo-Pacific	—
1.2.2	*Pilumnopeus vauquelini* (Audouin)	Alexandria (Balss, 1936) Haifa 1951 (Holthuis and Gottlieb, 1958)	Timsah (Keller, 1883) (Calman, 1927a; Monod, 1938; Holthuis, 1956; Por and Ferber, 1972)	Red Sea and Persian Gulf	= *Pilumnus*; = *Heteropanope* Ovigerous females in Suez Canal (Gurney, 1927a)
4.2.2	*Pilumnus hirsutus* Stimpson	Alexander (Balss, 1936)	— ? —	Red Sea	—
1.1.3	*Portunus pelagicus* (Linnaeus)	Alexandria, Mersa Matruch (Fox, 1924) Haifa 1926 (Fox, 1926; Steinitz, 1929) Lebanon, Syria (Gruvel, 1931) Iskenderun (Gruvel, 1928) Cyprus (Demetropoulos and Neocleous, 1969)	(Krukenberg, 1888 a) Kabret 1893; → Port Said 1902 (Fox, 1926)	Red Sea, Indo-Pacific	= *Lupa*; = *Neptunus*; Sicily (Ariani and Sarra, 1969)
4.3.2	*Synalpheus hululensis* Coutière	Apollonia 1961 (Lewinsohn and Holthuis, 1964)	— ? —	Red Sea, West Indian Ocean	—
2.2.3	*Thalamita poissoni* (Audouin)	Bat Yam 1952; Haifa 1956 (Holthuis and Gottlieb, 1958) Cyprus (Gilat, 1969)	Little Bitter Lake (Calman, 1927a)	Red Sea, Persian Gulf, Indo-Pacific	—
2.2.3.	*Trachypenaeus curvirostris* (Stimpson)	Haifa (Steinitz, 1929; Holthuis and Gottlieb, 1958) Iskenderun (Geldiay and Kocatas, 1972)	(Balss, 1927; Calman, 1927a; Gruvel, 1936; Monod, 1937; Holthuis, 1956)	Red Sea → Indian Ocean	= *Metapenaeus*; = *M. palestinensis*

	Mollusca: Amphineura				
4.2.2	*Chiton platei* Thiele	'Akko 1934 Caesarea 1971 (Barash, pers. comm.)	— ? —	Gulf of 'Aqaba (Leloup, 1960)	—
	Mollusca: Pelecypoda (Bivalvia)				
3.2.2	*Arca natalensis* Krauss	Yaffo, Haifa (Haas, 1937) Sirbonis (Bardawil Lagoon) → Qiryat Haim (Barash and Danin, 1972)	(Moazzo, 1939; Ryland, 1951)	Gulf of Suez, Aden Persian Gulf, Indo-Pacific	= *A. rufescens* Rare in Canal and along Israel shores
1.3.2	*Atactodea striata* (Gmelin)	Israel: Netanya 1973 (Barash, personal communication)	(Tillier and Bavay, 1905; Moazzo, 1939; Ryland, 1951; Barash and Danin, 1972)	Indo-Pacific	= *Mesodesma*
1.1.4	*Brachidontes (Hormomya) variabilis* (Krauss)	Atlit, Haifa (Haas, 1937) Lebanon (Gruvel and Moazzo, 1931)	Menzaleh, Timsah, Great Bitter Lake (Fuchs, 1878) Port Said (Keller, 1882) (Barash and Danin, 1972)	Gulf of Suez, Gulf of 'Aqaba, Aden, West Indian Ocean	= *Mytilus*; = *M. pharaonis* Syracuse, Sicily (Geronimo, 1971)
3.1.1	*Chama broderipi* Reeve	Alexandria (Tillier and Bavay, 1905; Pallary, 1912a)	(Moazzo, 1939)	Red Sea? Madagascar	—
3.2.2	*Clementia papyracea* (Gmelin)	Gaza (Haas, 1937) El Arish 1968 (Barash and Danin, 1972) "Rare" → Haifa	(Moazzo, 1939; Ryland, 1951)	Gulf of Suez Persian Gulf Indo-Pacific	= *C. cumingi*
3.3.2	*Gastrochaena (Cucurbitula) cymbium* Spengler	Israel: Many capsules 1954 onwards (Barash and Danin, 1972)	(Moazzo, 1939: "rather common")	Gulf of Suez Red Sea Indo-Pacific	= *Cucurbitula* = *G. deshayesi* Only capsules found along Israel coast (Barash and Danin, 1972)
1.3.2	*Laternula subrostrata* (Lamarck)	Ashdod 1961; shells 1958, widerspread (Barash and Danin, 1972)	(Sturany, 1899 → Ryland, 1951)	Gulf of Suez Persian Gulf Indo-Pacific	= *Anatina* = *A. anatina*

[a] See key to symbols in Table 5.

Table 2 (continued)

Category[a]	Taxon and species (species in alphabetical order)	Record of distribution and authority			Synonymy and Remarks
		Mediterranean	Suez Canal	Red Sea and Indo-Pacific	
1.2.2	*Mactra olorina* Philippi	Beirut (Gruvel and Moazzo, 1931) Sirbonis (Bardawil Lagoon) (Barash and Danin, 1972)	(Keller, 1895; Bavay, 1898 → Barash and Danin, 1972)	Gulf of Suez Gulf of 'Aqaba Persian Gulf Indo-Pacific	—
1.2.3	*Malleus (Parimalleus) regula* (Forskål)	'Akko, Tyre (Gruvel and Moazzo, 1931; Haas, 1937) Cyprus (Demetropoulos, 1971)	(Tillier and Bavay, 1905; widespread, Barash and Danin, 1972)	Gulf of Suez Indo-Pacific	—
4.3.2	*Modiolus arcuatulus* Hanley	Live: Sirbonis (Bardawil Lagoon) (Barash and Danin, 1972) Israel many shells 1964 →	— ? —	Red Sea, Aden Indo-Pacific	= *Arcuatula arcuatula* ? = *A. senhousia*
3.3.2	*Modiolus glaberrimus* (Dunker)	Israel 1960 → N. to Haifa (Barash and Danin, 1972)	(Moazzo, 1939)	Gulf of Suez Red Sea Indo-Pacific	= *M. perfragilis*
3.2.2	*Paphia textile* (Gmelin)	Haifa 1946 (Barash and Danin, 1972) Gaza, Haifa (Haas, 1948; Gilat, 1964; Barash and Danin, 1972)	(Moazzo, 1939; Ryland, 1951)	Gulf of Suez Red Sea Indo-Pacific	= *Tapes textrix* ? = *P. undulata*
1.3.2	*Papyridea papyracea* (Gmelin)	Haifa 1955 (Barash and Danin, 1972) Rather rare	(Bavay, 1898; → Ryland, 1951; Barash and Danin, 1972)	Gulf of Suez, Gulf of 'Aqaba Persian Gulf Indo-Pacific	= *Cardium* = *C. tenuicostatum*
	Mollusca: Gastropoda				
3.3.2	*Bursatella leachi savigniana* Audouin	Haifa 1955, Israel coast (Eales, 1970; Barash and Danin, 1972)	El Qantara 1967 (Barash and Danin, 1972)	Red Sea	= *Notarchus savignianus* Malta (Bebbington, 1970) S. Turkey (Swennen, 1961)

1.3.2	*Cellana rota* (Gmelin)	'Akko (Christiaens, 1967; Barash and Danin, 1972 – shells)	(Tillier and Bavay, 1905 → Barash and Danin, 1972)	Gulf of Suez Gulf of 'Aqaba Indo-Pacific	= *Helcioniscus* = *Patella* "Occasional migrant to Mediterranean" (Barash and Danin, 1972)
1.1.2	*Cerithium (Thericium) scabridum* Philippi	Alexandria → Jaffa (Pallary, 1912a, b) 'Akko (Haas, 1937) Lebanon, Syria (Pallary, 1938)	(Keller, 1882 along the Canal; → Barash and Danin, 1972)	Gulf of Suez, Persian Gulf Aden	= *Thericium*
4.3.2	*Cerithium (Vertagus) kochi* Philippi	Haifa 1963 → "enormous increase" (Barash and Danin, 1972)	— ? —	Gulf of Suez, Red Sea Indo-Pacific	= *Vertagus* ? = *C. recurvum* = *Ochetoclava* "One of the most abundant offshore species of Israel" (Barash and Danin, 1972)
1.2.2	*Diodora (Diodora) rueppelli* (Sowerby)	Haifa (Haas, 1948) Israel 1955 (Barash and Danin, 1972)	(Keller, 1882; Tillier and Bavay, 1905; → widespread, Barash and Danin, 1972)	Gulf of Suez Gulf of 'Aqaba West Indian Ocean	= *Fissurella*
1.2.2	*Fusinus marmoratus* Philippi	'Akko (Gruvel and Moazzo, 1931; Pallary, 1938)	Very common! (Keller, 1882; Bavay, 1898; → Barash and Danin, 1972)	Gulf of Suez Gulf of 'Aqaba Persian Gulf West Indian Ocean	= *Fusus* = *F. tuberculatus* "Occasional migrant to Mediterranean" (Barash and Danin, 1972)
4.3.2	*Isanda (Vanitrochus)* cf. *holdsworthiana* (Nevill)	Haifa 1966 (Barash and Danin, 1972)	— ? —	Gulf of Suez Persian Gulf, Ceylon	The genus is Indo-Pacific
1.2.2	*Murex (Murex) tribulus* Linnaeus	Lebanon (Pallary, 1938, 1 shell) Live: El Arish, Yunis, 1954 (Barash and Danin, 1972)	(Keller, 1882; Bavay, 1898; → Barash and Danin, 1972)	Gulf of Suez Gulf of 'Aqaba Indo-Pacific	= *M. crassispina*

[a] See key to symbols in Table 5.

Table 2 (continued)

Category[a]	Taxon and species (species in alphabetical order)	Record of distribution and authority			Synonymy and Remarks
		Mediterranean	Suez Canal	Red Sea and Indo-Pacific	
2.3.3	*Nerita sanguinolenta* Menke	Karpathos (Nordsieck, 1972, sine Mienis, 1974)	(Tomlin, 1927; → Barash and Danin, 1972)	Red Sea	= *Neritina kinzelbachi* = *N. forskalii* Not reported from Levant Basin
3.3.2	*Notarchus indicus* Schweigger	Ashdod (Eales, 1970; Barash and Danin, 1972)	(Gruvel, 1936)	Gulf of 'Aqaba Indo-Pacific	—
4.3.2	*Rissoina (Rissolina) bertholleti* Issel	Shells on Israel coast 1965 (Barash and Danin, 1972)	— ? —	Gulf of Suez, Aden Persian Gulf West Indian Ocean	"Shells not uncommon" (Barash and Danin, 1972)
1.3.2	*Siphonaria kurracheensis* Reeve	Shiqmona, Akhziv 1965 → (Barash and Danin, 1972)	(Keller, 1895; Tomlin, 1927; Barash and Danin, 1972)	Gulf of Suez Gulf of 'Aqaba Indo-Pacific	= *S. laciniosa* Common in the Canal
2.2.2	*Thais carinifera* (Lamarck)	Tel Aviv 1956 Sirbonis (Bardawil Lagoon) → 'Akko	(Tomlin, 1927; Ryland, 1951; Barash and Danin, 1972)	Red Sea Persian Gulf Indo-Pacific	Egg capsules found in Mediterranean
	Bryozoa				
2.2.1	*Buskia setigera* Hincks	Port Said, Menzaleh Lock basin (Hastings, 1927)	All the Canal (Hastings, 1927)	Port Taufiq (Hastings, 1927) Indian Ocean, Malaya	—
4.3.2	*Hippaliosina acutirostris* Canu et Bassler	Haifa Bay (Powell, 1969a)	— ? —	S. Red Sea (Powell, 1967) Indo-Pacific	—
4.3.2	*Hippopodina feegeensis* (Busk)	Israel (Powell, 1969a) Ashdod, Caesarea (Eitan, 1974)	— ? —	Ghardaqa (Harmer, 1957) Indo-Pacific	Also W. Atlantic — but: "it represents the Indo-Pacific variety" (Eitan, 1974)
2.2.1	*Scrupocellaria jolloisii* (Audouin)	Port Said, outside E. jetty (Hastings, 1927)	All the Canal (Hastings, 1927)	Red Sea Indian Ocean	—

	Echinodermata[b]				
2.3.2	*Asterina wega* Perrier	Haifa, Beirut (Tortonese, 1966) Akhziv (Achituv, 1969)	Great Bitter Lake (Mortensen, 1926)	Gulf of Suez Gulf of 'Aqaba Persian Gulf Indo-Pacific	= *A. burtoni* "partim"
4.2.2	*Amphioplus laevis* Lyman	Rafiah, 1947 (Tortonese, 1953–54)	— ? —	Red Sea (Burfield, 1924) Persian Gulf (Mortensen, 1940) Indo-Pacific	First record of the genus in the Mediterranean
	Tunicata: Ascidiacea				
2.2.2	*Ascidia cannelata* (Savigny-Oken)	Israel coast (Pérès, 1958a)	Lake Timsah → Km 76 (Harant, 1927)	Gulf of Suez (Hartmeyer, 1915) Indo-Pacific	—
2.1.2	*Herdmania momus* (Savigny)	Alexandria (Harant, 1939) Israel Med. Haifa (Pérès, 1958a, b)	All the Canal (Harant, 1927)	Port Taufiq (Harant, 1927)	= *Pyura* "Extrèmement abondante. Sans doute immigré de la Mer Rouge." (Pérès, 1958a)
2.2.2	*Phallusia nigra* Savigny	Israel: Caesarea 1952 (Pérès, 1958a)	All the Canal (Harant, 1927) Lake Timsah (Tortonese, 1952)	Gulf of Suez (Harant, 1927; Hartmeyer, 1915)	—
2.2.2	*Symplegma viride* Herdman	Israel: Caesarea 1951 (Pérès, 1958a)	Km 46, Km 64 (Harant, 1927)	Red Sea ("commune": Pérès, 1958a)	—
	Pisces				
1.2.2	*Apogonichthyoides nigripinnis* (Cuvier et Valenciennes)	Mediterranean coast of Israel (Haas and Steinitz, 1947; → Ben-Tuvia, 1963)	(Tillier, 1902) Kabret, Port Said (Norman, 1927) (Tortonese, 1948)	Red Sea Indian Ocean	= *Apogon* = *A. thurstoni* = *A. taeniatus* = *A. bifasciatus*
1.1.2	*Atule djeddaba* (Forsskål)	Israel (W. Steinitz, 1927; Liebmann, 1934; Ben-Tuvia, 1963)	Bitter Lake (Tillier, 1902) Kabret (Norman, 1927) (Chabanaud, 1934) Timsah (Tortonese, 1948)	Red Sea Indo-Pacific	= *Scomber* = *Caranx* = *C. calla*

[a] See key to symbols in Table 5.
[b] *Note added in proof:* Tom (1976) reports also the brittle star *Ophiactis prava* (Mortensen) from Haifa.

Table 2 (continued)

Category[a]	Taxon and species (species in alphabetical order)	Record of distribution and authority			Synonymy and Remarks
		Mediterranean	Suez Canal	Red Sea and Indo-Pacific	
2.2.2	*Callionymus filamentosus* Cuvier et Valenciennes	Israel (Tortonese, 1953; Ben-Tuvia, 1953, 1963)	Timsah (Norman, 1929) Great Bitter Lake (Chabanaud, 1932)	Red Sea Indo-Pacific	= *Calliurichthys* = *C. haifae*
1.3.1	*Crenidens crenidens* (Forsskål)	Bardawil Lagoon (Lourie and Ben-Tuvia, 1970)	All the Canal (Tillier, 1902) Great Bitter Lake (Ben-Tuvia, 1975b)	Gulf of Suez, El Bilaiyim lagoon Indo-Pacific	—
3.2.2	*Dollfusichthys sinusarabici* Chabanaud	Israel (Ben-Tuvia, 1953) Haifa (Chabanaud, 1934)	Great Bitter Lake (Chabanaud, 1932)	Gulf of Suez, Red Sea	= *Cynoglossus* —
1.2.3	***Dussumieria acuta*** (Valenciennes)	Israel (Lissner, 1949; Ben-Tuvia, 1953) Iskenderun, Mersin (Ben-Tuvia, 1953)	(Tillier, 1902; Chabanaud, 1932; Bertin, 1943; Tortonese, 1948)	Red Sea Indo-Pacific	= *D. productissima* —
4.3.2	*Epinephelus tauvina* (Forsskål)	Haifa (Ben-Tuvia and Lourie, 1969)	— ? —	Red Sea Persian Gulf Indo-Pacific	—
1.2.3	*Hemiramphus far* (Forsskål)	Israel (Hornell, 1935; Haas and Steinitz, 1947; Ben-Tuvia, 1963) Iskenderun (Kosswig, 1950) Rhodes (Tortonese, 1947)	Southern Canal (Tillier, 1902) Kabret (Norman, 1927)	Red Sea Indo-Pacific	—
2.3.1	*Herclotsichthys punctatus* (Rüppell)	Bardawil Lagoon (Ben-Tuvia, 1975b)	→ Timsah (Norman, 1927)	Red Sea	—
1.3.3	*Himantura uarnak* (Forsskål)	Israel (Ben-Tuvia, 1953, 1963) Mersin (Ben-Tuvia, 1966)	(Tillier, 1902) Great Bitter Lake (Gruvel, 1936)	Red Sea Indo-Pacific	= *Raja* = *Trygon* = *Dasyatis*

3.2.4	*Holocentrus ruber* (Forsskål)	Israel (Haas and Steinitz, 1947; Ben-Tuvia, 1953, 1963) Rhodes (Laskaridis, 1948b) Iskenderun (Kosswig, 1950) Cyprus (Demetropoulos and Neocleous, 1969) Tobruk (Stirn, 1973)	? Suez harbour (Gruvel and Chabanaud, 1937)	Red Sea Indo-Pacific	= *Sciaena*
1.2.4	*Leiognathus klunzingeri* (Steindachner)	N. of Port Said (Norman, 1929) Syria (Gruvel, 1931) Israel (Liebmann, 1934) Iskenderun, Rhodes (Erazi, 1943) Lampedusa (Ben-Tuvia, 1966) Cyprus (Demetropoulos and Neocleous, 1969)	→ Port Said (Tillier, 1902; Tortonese, 1948)	Red Sea	= *Equula* = *L. oblongus* = *L. lineolatus* = *L. mediterraneus*
1.2.1	*Liza carinata* (Valenciennes)	W. of Port Said (Wimpenny, 1931) E. Mediterranean (Norman, 1929)	(Tillier, 1902; Tortonese, 1948)	Red Sea Indian Ocean	= *Mugil* = *M. seheli*
4.3.2	*Pelates quadrilineatus* (Bloch)	Haifa (Lourie and Ben-Tuvia, 1970)	— ? —	Gulf of Suez Massawa, Red Sea Indo-Pacific	—
1.2.2	*Platycephalus indicus* (Linnaeus)	Israel (Haas and Steinitz, 1947; Ben-Tuvia, 1966: "single specimens". Egypt (Krefft, 1963)	Southern Canal (Keller, 1882) → Port Said (Tillier, 1902; Norman, 1927)	Red Sea Indo-Pacific	= *Callionymus* = *P. insidiator*
1.1.4	*Pranesus pinguis* (Lacépède)	Alexandria – Port Said (Tillier, 1902) Port Said (Jordan and Hubbs, 1917) Haifa (Norman, 1927) Cyprus (Norman, 1929) Mersa Matruch (Norman, 1929) Iskenderun (Fowler and Steinitz, 1956)	(Tillier, 1902: "Passant d'une mer a l'autre")	Red Sea Indo-Pacific	= *Atherina* = *A. forskalii* = *Hepsetia*

[a] See key to symbols in Table 5.

Table 2 (continued)

Category[a]	Taxon and species (species in alphabetical order)	Record of distribution and authority			Synonymy and Remarks
		Mediterranean	Suez Canal	Red Sea and Indo-Pacific	
4.3.2	*Rastrelliger kanagurta* (Cuvier)	Israel, 1967 (Collette, 1970)	— ? —	Indo-Pacific	—
4.2.3	*Saurida undosquamis* (Richardson)	Israel (Ben-Tuvia, 1953) Mersin 1956 (Ben-Tuvia, 1966) Cyprus (Ben-Tuvia, 1962) Izmir (Artuz, fide Ben-Tuvia, 1973)	— ? —	Red Sea Indo-Pacific	= ? *S. grandisquamis* Port Said (Ben-Tuvia, 1966) Suez Canal (Ben-Tuvia, 1966, 1972)
4.3.2	*Scomberomorus commerson* (Lacépède)	Israel (Ben-Tuvia, 1971a) Lebanon (George and Athanassiou, 1965)	— ? —	Indo-Pacific	—
1.3.4	*Siganus luridus* (Rüppell)	Israel 1955 (Ben-Tuvia, 1963) Rhodes (Ben-Tuvia, 1972) Gulf of Tunis (Ktari-Chakroun and Bahloul, 1971) Izmir (Artuz, fide Ben-Tuvia, 1973) Cyprus (Demetropoulos and Neocleous, 1969) Tobruk (Stirn, 1973)	(Tillier, 1902: "Lagunes de Suez")	Red Sea	= *Amphacanthus* (Tillier, 1902: "Jamais vue au dehors de la Baie de Suez")
1.1.4	*Siganus rivulatus* (Forsskål)	Israel (W. Steinitz, 1927; Liebmann, 1934) Syria (Gruvel, 1929) Cyprus (Norman, 1929) "Egean" (Brunelli and Bini, 1934) Rhodes (Tortonese, 1947) Iskenderun (Haas and Steinitz, 1947) Tobruk (Stirn, 1973)	→ Bitter Lake (Tillier, 1902) Kabret (Norman, 1927) (Tortonese, 1948)	Red Sea W. Indian Ocean	= *Scarus* = *Teuthis* = *Amphacanthus* = *S. siganus* = *S. nebulosus* = *S. spinus*

1.3.3	*Sphaeroides spadiceus* (Richardson)	Iskenderun (Kosswig, 1950) Samos (Anandiades, 1952) Israel (Ben-Tuvia, 1953) Mersin (Ben-Tuvia, 1966)	(Tillier, 1902: "Assez abondante dans le Petit Lac Amer en 1890"; Gruvel and Chabanaud, 1937)	Gulf of Suez, Red Sea Indo-Pacific	= *Lagocephalus* = *Tetrodon* = *T. lunaris* "Rare along the coast of Israel but often caught" (Ben-Tuvia, 1966)
3.3.4	*Sphyraena chrysotaenia* Klunzinger	Israel (Ben-Tuvia, 1953) Beirut, Mersin (Ben-Tuvia, 1966) Cyprus (Demetropoulos and Neocleous, 1969) Tobruk (Stirn, 1973)	→ Timsah (Chabanaud, 1933b) (Gruvel and Chabanaud, 1937)	Gulf of Suez, Red Sea Indo-Pacific	= *S. obtusata*
1.1.4	*Stephanolepis diaspros* Fraser-Brunner	Israel (W. Steinitz, 1927; Liebmann, 1934) Cyprus (Hornell, 1935) Rhodes (Laskaridis, 1948b) Iskenderun (Kosswig, 1950) Taranto (Tortonese, 1973a)	(Tillier, 1902 → Fraser-Brunner, 1940)	Gulf of Suez, Red Sea Persian Gulf	= *Monacanthus setifer* = *S. hispidus* = *S. ocheticus* = *S. weberi*
4.3.2	*Tylosurus choram* (Rüppell)	Lebanon (George et al., 1964)	— ? —	Indo-Pacific	—
3.3.2	*Upeneus asymmetricus* Lachner	Israel (Ben-Tuvia, 1966) Lebanon (George and Athanassiou, 1966)	— ? —	Red Sea Indo-Pacific	= *U. tragula*
3.2.4	*Upeneus moluccensis* Blecker	Israel (Haas and Steinitz, 1947) Rhodes (Laskaridis, 1948a) Iskenderun (Kosswig, 1950) Izmir (Artuz, fide Ben-Tuvia, 1973) Cyprus (Demetropoulos and Neocleous, 1969)	(Gruvel, 1936)	North Red Sea Indo-Pacific	= *Mulloides auriflama* = *M. flavolineata* Tobruk: "Probably *U. moluccensis*" (Stirn, 1973)

[a] See key to symbols in Table 5.

Table 3. Low probability Lessepsian migrants (LPLM)

Category[a]	Taxon and species (species in alphabetical order)	Record of distribution and authority			Remarks
		Mediterranean	Suez Canal	Red Sea and Indo-West-Pacific	
	Algae				
4.2.1	*Acrochaetium sargassicola* Borg.	Port Said (Aleem, 1950)	—?—	"agrees well" with Bombay specimen	—
4.2.1	*Acrochaetium subseriatum* Borg.	Port Said, El Arish (Aleem, 1950)	—?—	India, Mauritius	—
4.2.2	*Caulerpa mexicana* Sonder ex Kützing	Israel → Syria (Rayss, 1941, 1955; Rayss and Edelstein, 1960)	—?—	Red Sea: Qoseir, Yenbo, Jiddah	Florida, Canaries ("Pantropicophyte": Rayss, 1955) ("migrant": Lipkin, 1972a)
2.2.4	*Caulerpa racemosa* (Forsskål) Agardh	Tunis (Hamel, 1930) Israel, Syria (Rayss, 1959) Asia Minor (Huvé, 1957)	Bitter Lake (Lami, 1932)	Red Sea: widespread	("can be described as pantropical": Aleem, 1950)
4.1.2	*Caulerpa scalpelliformis* (Brown) Agardh	Syria (Hamel, 1930) Israel (Carmin, 1934)	—?—	Gulf of Suez → Indo-West-Pacific	Also Brasilia, Canaries, Delagoa Bay
4.2.4	*Gracilaria arcuata* Zanardini	Tunis (Feldmann, 1931) Egypt (Aleem, 1948)	—?—	Red Sea → Indo-West-Pacific	("worldwide in warm seas": Aleem, 1948)
4.2.1	*Hypnaea cornuta* (Lamour) Agardh	Alexandria (Aleem, 1948)	—?—	Indo-West-Pacific	("worldwide": Aleem, 1948)
4.1.1	*Hypnaea nidifica* Agardh	Rhodes (Forti, 1928)	—?—	Thor? (Reinbold, 1898) Red Sea	—
4.2.1	*Kylinia spathoglossi* (Borg.) Aleem	Port Said, El Arish (Aleem, 1950)	—?—	South India	—
—	*Lophocladia lallemandii* (Mont.) Schmitz	Algier (Feldmann and Feldmann, 1938)	—?—	Red Sea: Suakin, Dahlak	("hypothetic migrant since not reported from Atlantic: Aleem, 1948)
4.3.2	*Padina gymnospora* (Kütz.) Vickers	Israel (Ramon and Friedman, 1966)	—?—	India?	—

	Porifera				
4.2.1	*Callyspongia viridis* (Keller)	Alexandria (Burton, 1936)	— ? —	Red Sea	("possible misidentification": Tsurnamal, pers. comm.)
4.1.1	*Mycale erythraeana* (Row)	Port Said (between the piers) (Burton, 1926)	— ? —	Red Sea	—
	Coelenterata: Hydroidea				
2.2.1	*Pennaria disticha australis* Bale	Alexandria (Billard, 1936)	Port Taufiq, Qantara (Billard, 1927)	Red Sea? Indo-West-Pacific	Also East Africa?
—	*Euphysora bigelowi* Maas	Egypt coast (Schmidt, 1972)	— ? —	Indian Ocean	—
—	*Laodicea fijiana* Agassiz et Mayer	Cyprus (Schmidt, 1972)	— ? —	Fiji?	—
	Ctenophora				
—	*Coeloplana* sp.	Israel (Haas, 1942)	— ? —		Indo-Pacific genus
	Annelida: Polychaeta				
4.3.2	*Asychis gotoi* Izuka	Israel, Sinai (Ben-Eliahu, 1970, 1972d)	— ? —	no Red Sea report Indo-West-Pacific	Found at Rovigno (Adria) (Fauvel, 1940)
—	*Chrysopetalum* gen. sp.?	Beyrouth (Laubier, 1966) Sinai? (Ben-Eliahu, 1970)	— ? —		("specimens at Elat and Baleares": Laubier, 1966)
4.3.2	*Cirriformia ancylochaeta* (Schmarda)	Beyrouth (Laubier, 1966)	— ? —		("parait être ... indopacifique": Laubier, 1966) (in W. Mediterranean? Bellan, 1964)
2.3.2	*Dasychone cingulata* (Grube)	Israel (Fauvel, 1955) Beyrouth (Laubier, 1966)	— ? —	Indo-West-Pacific	("separation from Mediterranean *D. lucullana* still unclear": Laubier, 1966)
2.3.2	*Eunice indica* Kinberg	Israel, Sinai (Ben-Eliahu, 1972d)	→ Qantara (Fauvel, 1927) Qantara (Ben-Eliahu, 1972d)		("possible confusion with Atlantic *E. vittata*. *E. indica* also tropical Atlantic", Ben-Eliahu, 1972d)

[a] See key to symbols in Table 5.

Table 3 (continued)

Category[a]	Taxon and species (species in alphabetical order)	Record of distribution and authority			Remarks
		Mediterranean	Suez Canal	Red Sea and Indo-West-Pacific	
2.3.2	*Leonnates decipiens* Fauvel	Sinai? (Ben-Eliahu, 1970)	Timsah (Fauvel, 1927)	Gulf of Suez (Fauvel, 1933) Indo-West-Pacific	—
4.3.3	*Lysidice collaris* Grube	Cyprus (Ben-Eliahu, 1972b)	—?—	Gulf of Aqaba (Elat) common (Ben-Eliahu, 1972d)	—
4.3.3	*Naineris quadriticeps* Day	Santorin? (Harmelin, 1969)	—?—	Red Sea	Two incomplete fragments tentatively identified
4.3.3	*Notomastus aberans* Day	Crete, Santorin (Harmelin, 1969)	—?—	Indo-West-Pacific	(Also "région marseillaise": Harmelin, 1969)
4.2.2	*Pseudeurythoë* cf. *acarunculata* Monro	Beyrouth (Laubier, 1966)	—?—	Red Sea?	(identification uncertain but genus mostly Indo-Pacific: Laubier, 1966)
4.3.3	*Rhodine loveni* Malmgren	Israel (Fauvel, 1957) Sinai; Cyprus (Ben-Eliahu, 1970)	—?—	Red Sea? Indo-West-Pacific	W. Mediterranean (Reyss, 1964)
4.3.3	*Scololops chevalieri candiensis* Harmelin	Crete (Harmelin, 1969)	—?—	—	Nominate species: Red Sea and Persian Gulf
2.3.2	*Spirobranchus giganteus coutierei* Gravier	Beyrouth (Laubier, 1966) Israel? (Ben-Eliahu, 1972d)	(Potts, 1928; Ben-Eliahu, 1972d)	Red Sea	("pantropical species": Laubier, 1966)
4.3.3	*Tharyx dorsobranchialis* Kirkegaard	Crete (Harmelin, 1969) Sinai; Cyprus (Ben-Eliahu, 1970)	—?—	Red Sea	South Africa; West Africa
	Crustacea: Copepoda				
1.1.1	*Pseudodiaptomus salinus* Giesbrecht	Port Said in the Mediterranean (Thompson and Scott, 1903)	All the Canal	Red Sea Indo-West-Pacific	—

3.3.2	*Scottolana bulbosa* (Por)	Israel (Por, 1964)	El Qantara (Por and Marcus, 1972)	—	—
	Crustacea: Tanaidacea				
4.3.2	*Apseudes intermedius* Hansen	Alexandria (Larwood, 1940) Israel (Băcescu, 1961)	— ? —	Red Sea Indo-West-Pacific	Cape Verde Morocco
	Crustacea: Amphipoda				
2.1.2	*Elasmopus pectenicrus* (Bate)	Alexandria (Schellenberg, 1936) Eastern Mediterranean (Ruffo, 1959) Caesarea, Haifa (Gottlieb, 1960)	"near Port Said" (Schellenberg, 1928)	Red Sea, but not Gulfs of Aqaba and Suez	Atlantic Fouling organism
—	*Stenothoë gallensis* Walker	Alexandria (Schellenberg, 1936)	— ? —	Indian Ocean	Also Marseilles "Harbour organism"
	Crustacea: Cumacea				
—	*Eocuma sarsii* (Kossman)	Alexandria (Steuer, 1938)	— ? —	Red Sea Indian Ocean	Also from Cannes (Western Mediterranean)
4.3.2	*Iphinoë crassipes haifae* Băcescu	Israel (Bacescu, 1961)	— ? —	—	Nominate species: Red Sea
	Crustacea: Isopoda				
4.2.1	*Apanthura sandalensis* Stebbing	Alexandria (Larwood, 1940)	— ? —	Red Sea India	—
	Crustacea: Decapoda				
4.2.2	*Alpheus inopinatus* Holthuis et Gottlieb	Israel (Holthuis and Gottlieb, 1958)	— ?—	Pakistan East Africa (Holthuis, pers. comm.)	("First of the *edwardsi* group in Mediterranean": Holthuis and Gottlieb, 1958)

[a] See key to symbols in Table 5.

Table 3 (continued)

Category[a]	Taxon and species (species in alphabetical order)	Record of distribution and authority: Mediterranean	Suez Canal	Red Sea and Indo-West-Pacific	Remarks
4.3.2	*Alpheus rapacida* de Man	Israel (Lewinsohn and Holthuis, 1964)	— ? —	Indo-West-Pacific, not Red Sea	—
—	*Automate branchialis* Holthuis et Gottlieb	Israel (Holthuis and Gottlieb, 1958)	— ? —	—	("First *Automate* in Mediterranean": Holthuis and Gottlieb, 1958) ("might have come through Suez Canal, but this is not certain": Holthuis, personal communication)
4.3.3	*Ixa monodi* Holthuis et Gottlieb	Mersin Bay (Holthuis and Gottlieb, 1956)	— ? —	Gulf of Suez	Indo-Pacific genus
4.2.2	*Leptochela pugnax* de Man	Israel (Holthuis and Gottlieb, 1958)	— ? —	not Red Sea Indo-West-Pacific	—
2.3.2	*Lucifer hanseni* Nobili	Port Said (Balss, 1927)	Abundant in Suez Canal	Red Sea Indo-West-Pacific	—
4.2.2	*Ogyrides mjöbergi* (Balss)	Israel (Holthuis and Gottlieb, 1958)	— ? —	not Red Sea Indo-West-Pacific	("might have been overlooked": Holthuis and Gottlieb, 1958)
2.2.1	*Periclimenes calmani* Tattersall	Port Said (Balss, 1927; Gurney, 1927a; Fox, 1929)	Bitter Lake (Gurney, 1927a; Holthuis, 1956)	Sudan Malaya	—
	Mollusca: Pelecypoda (Bivalvia)				
1.2.1	*Chama* cf. *cornucopiae* Reeve	Israel, 1 shell (Barash and Danin, 1972)	Suez Canal, widespread	Gulf of Suez Gulf of Aqaba Aden	—
1.2.1	*Gafrarium pectinatum* (L.)	El Arish, 1 valve (Barash and Danin, 1972)	Suez Canal, widespread	Gulf of Suez Gulf of Aqaba Indo-West-Pacific	—

1.2.2	*Modiolus auriculatus* Krauss	Israel, shells on beach (Haas, 1937)	Suez Canal, widespread	Gulf of Suez Gulf of Aqaba Indo-West-Pacific	—
4.2.2	*Papyridea australe* (Sowerby)	Israel, 3 specimens (Haas, 1948)	— ? —	Gulf of Suez Aden Persian Gulf Indo-West-Pacific	—
1.1.4	*Pinctada radiata* (Leach) = *Pteria occa* Reeve	Tunis 1895 (Bavay, 1897) Cyprus, Jaffa (Monterosato, 1899) Israel, widespread	All the Canal (Keller, 1888)	Gulf of Suez (Vaillant, 1865) Indo-West-Pacific Brazil, Bermuda	("peu probable que l'espèce . . . a pu parcourir près de 3000 km . . . en trente ans": Monterosato, 1899)
—	*Spondylus spectrum* Reeve	Israel, 1 shell (Aharoni, 1934)	— ?—	Red Sea Indo-West-Pacific	—
	Mollusca: Gastropoda				
4.3.2	*Alvania orbignyi* (Audouin)	Israel, few shells (Barash and Danin, 1972)	— ? —	Gulf of Suez Red Sea	—
2.3.2	*Anachis savignyi* Moazzo	Israel, shells (Mienis, 1972)	(2 shells: Tomlin, 1927)	Gulf of Suez	—
1.2.2	*Cerithium erythraeonense* Lamarck	Israel, 1 shell (Haas, 1937)	(Sturany, 1903; Tomlin, 1927; Moazzo, 1939)	Gulf of Suez Indo-West-Pacific	—
4.2.2	*Conus (Puncticulis) arenatus* Hwass	Israel, 1 shell (Haas, 1937)	— ? —	Gulf of Suez Indo-West-Pacific	("Incidental" in Medit. Barash and Danin, 1972)
4.2.4	*Erronea caurica* (L.)	Greece, 1932 (Settepassi, 1968) Rhodes (Spada, 1970)	— ? —	Gulf of Suez Indo-West-Pacific	—
4.2.2	*Haliotis pustulata* Reeve	Lebanon? (Gohar, 1954) Israel, 1 shell (Talmadge, 1971)	— ? —	Gulf of Suez Indo-West-Pacific	—
4.2.4	*Lophiotoma indica* (Roding)	Alexandria, 1 specimen (Steuer, 1939b) Palermo (Parenzan, 1970)	— ? —	Indo-West-Pacific *not* Red Sea	("Misidentification": Mienis, pers. comm.)
1.2.2	*Pirenella cailliaudi* (Potiez et Michaud)	Syria (Pallary, 1938) southern Israel (Barash and Danin, 1972)	(Sturany, 1903)	Gulf of Suez Western Indian Ocean	("Isthmic fauna": Por, 1971a)

[a] See key to symbols in Table 5.

Table 3 (continued)

Category[a]	Taxon and species (species in alphabetical order)	Record of distribution and authority			Remarks
		Mediterranean	Suez Canal	Red Sea and Indo-West-Pacific	
4.3.4	*Quoyula madreporarum* Sowerby	3 shells near Bari (Ghisotti, 1968)	— ? —	Gulf of Suez Indo-West-Pacific	Euryoek species with wide spreading potential (Barash and Danin, 1972)
1.3.2	*Trochus erythraeus* Brocchi	Israel, 1 shell (Barash and Danin, 1972)	(Tillier and Bavay, 1905; Barash and Danin, 1972)	Gulf of Suez Western Indian Ocean	—
4.3.4	*Umbonium* cf. *vestiarium* (L.)	Libya, several specimens (Spada, 1970)	— ? —	Gulf of Suez Indo-West-Pacific	—
	Bryozoa				
2.2.2	*Celleporaria aperta* (Hincks)	Port Said (Hastings, 1927) S. Israel (Powell, 1969a)	(Hastings, 1927; Eitan, 1972)	Red Sea Indo-West-Pacific	Also warm Atlantic
4.1.1	*Membranipora savartii* Audouin	Port Said (Hastings, 1927)	— ? —	Red Sea Indo-West-Pacific	Also warm Atlantic
4.3.2	*Thalamoporella gothica indica* (Hincks)	Israel (Powell, 1969a)	— ? —	Red Sea? N. Indian Ocean	Nominate species: amphiatlantic
	Echinodermata				
2.2.2	*Ophiactis savignyi* Müller and Troschel	Port Said (Mortensen, 1926) Alexandria (Mortensen and Steuer, 1937) Israel (Tortonese, 1954) Lebanon (Tortonese, 1966)	(Mortensen, 1926; Por and Ferber, 1972)	Red Sea Indo-West-Pacific	("doubtful": Tortonese, 1973b) Banyuls (Guille, 1968)
	Tunicata: Ascidiacea				
2.2.1	*Ecteinascidia moorei* Herdmann	Mediterranean (Harant, 1927)	Kabret (Harant, 1927)	Red Sea	("cosmopolitan": Harant, 1927)

2.2.1	*Eusynstyela hartmeyeri* Michaud	Port Said harbour (Harant, 1927)	All the Canal (Harant, 1927)	Gulf of Suez	—
4.2.4	*Metrocarpa nigrum* (Herdmann)	Tunis (Pérès, 1954) Israel (Pérès, 1958a)	— ? —	Gulf of Suez Red Sea	("senegalian relic": Pérès, 1958a)
	Enteropneusta				
2.2.1	*Saccoglossus gurneyi* Robinson	Alexandria (Steuer, 1939a)	All the Canal (Robinson, 1927b)	— ? —	—
	Pisces				
4.3.4	*Abudefduf saxatilis* L.	Naples (Tardent, 1959)	— ? —	Red Sea Indo-West-Pacific	("circumtropical floatsam species" Briggs, 1974b)
1.2.2	*Aphanius dispar* (Rüppell)	Israel (Mendelssohn, 1947)	(Tillier, 1902)	Red Sea West Indian Ocean	("Isthmic" element: Por, 1971a)
4.3.3	*Sebastapistes nuchalis* (Günther)	Cyprus (Froiland, 1972)	— ? —	East Africa Indo-West-Pacific	—
	Mammalia				
—	*Dugong dugon* Müller	Israel, 1 specimen (Aharoni, 1930)	— ? —	Gulf of Suez	("misidentification": Mendelssohn, pers.comm.)

[a] See key to symbols in Table 5.

Table 4. Rejected cases of Lessepsian migrants (only first reference as migrant is given)

Porifera
Didiscus placospongoides (Burton, 1926)

Hydrozoa
Filellum serratum (Steinitz, 1967)

Polychaeta
Eurythoë complanata (Fauvel, 1955)
Opisthosyllis brunnea (Steinitz, 1967)

Sipuncula
Aspidosiphon elegans (Wesenberg-Lund, 1957a)

Crustacea
Canthocalanus pauper (Thompson and Scott, 1903)
Gammarus foxi (Schellenberg, 1936)
Gonodactylus falcatus (Heller, 1863)
Neptunus sanguinolentus (Babič, 1911)
Parathelges racovitzai (Codreanu, 1967)
Plagusia tuberculata (Stiasny, 1908)
Platimaia wywillethompsoni (Pesta, 1918)
Temora discaudata (Thompson and Scott, 1903)
Thalamita admete (Guérin-Meneville, 1832)
Thenus orientalis (Babič, 1911)

Mollusca
Cypraea annulus (Fischer, 1870)
Cypraea erosa (Fischer, 1870)
Cypraea lynx (Fischer, 1870)
Hippopus hippopus (Haas, 1937)
Hydrobia musaensis (Carus, 1893)
Murex ponderosus (Steinitz, 1967)
Strombus lentiginosus (Aharoni, 1934).
Turris indica (Steuer, 1939b)

Echinodermata
Astropecten bonnieri (Russo, 1935)
Astropecten inutilis (Russo, 1935)
Astropecten nobilis (Russo, 1935)
Astropecten polyacanthus (Russo, 1935)
Heterocentrotus mammillatus (Gauthier, 1874)
Prionocidaris baculosa (Russo, 1935)

Pisces
Carcharhinus brevipinna (Ben-Tuvia, 1966)
Etrumeus teres (Whitehead, 1963)
Hemiramphus marginatus (Gohar, 1954)
Parexocoetus mento (Bruun, 1935)

3.2 The Levant Basin of the Mediterranean

In 1908 the Danish Oceanographic Expedition on the "Thor" worked in the Mediterranean. Two of their findings are directly relevant to our subject. Schmidt (1932) proposed dividing the Mediterranean into two basins, the Western and the Eastern Mediterranean, separated by the sill between Sicily and Tunisia. Thomsen (1931) established for the first time that the Mediterranean is excessively poor in nutrients and that this depletion increases from west to east.

The Eastern Mediterranean is usually defined as a very heterogenous oceanic body (Fig. 29). It contains the Adriatic Sea and the Aegean Sea, each with a peculiar history and a particular hydrography. In these two basins the temperatures and salinities are mainly lower than those of the Mediterranean. Along the African coast there is an inshore sea area between the two opposing Gulfs of Sidra and Gabes, confined between the west coast of Cyrenaica and the east coast of Tunisia; here salinities are of the West Mediterranean type, i.e. below 39‰, but temperatures are high and subtropical and the tides are very pronounced. The Ionian Sea, north of the "Sidra Sea" and south of the Adriatic, bordered to the east by Crete and to the west by Sicily, is in fact a transitional body of water without any peculiar characteristics.

Finally there is the Sea of Levant, or the Levant Basin, delineated by the South Anatolian coast, the Levant shores and the opposite Egyptian and north Sinai coasts. The western limits of the Levant Basin should, in my view, be defined by

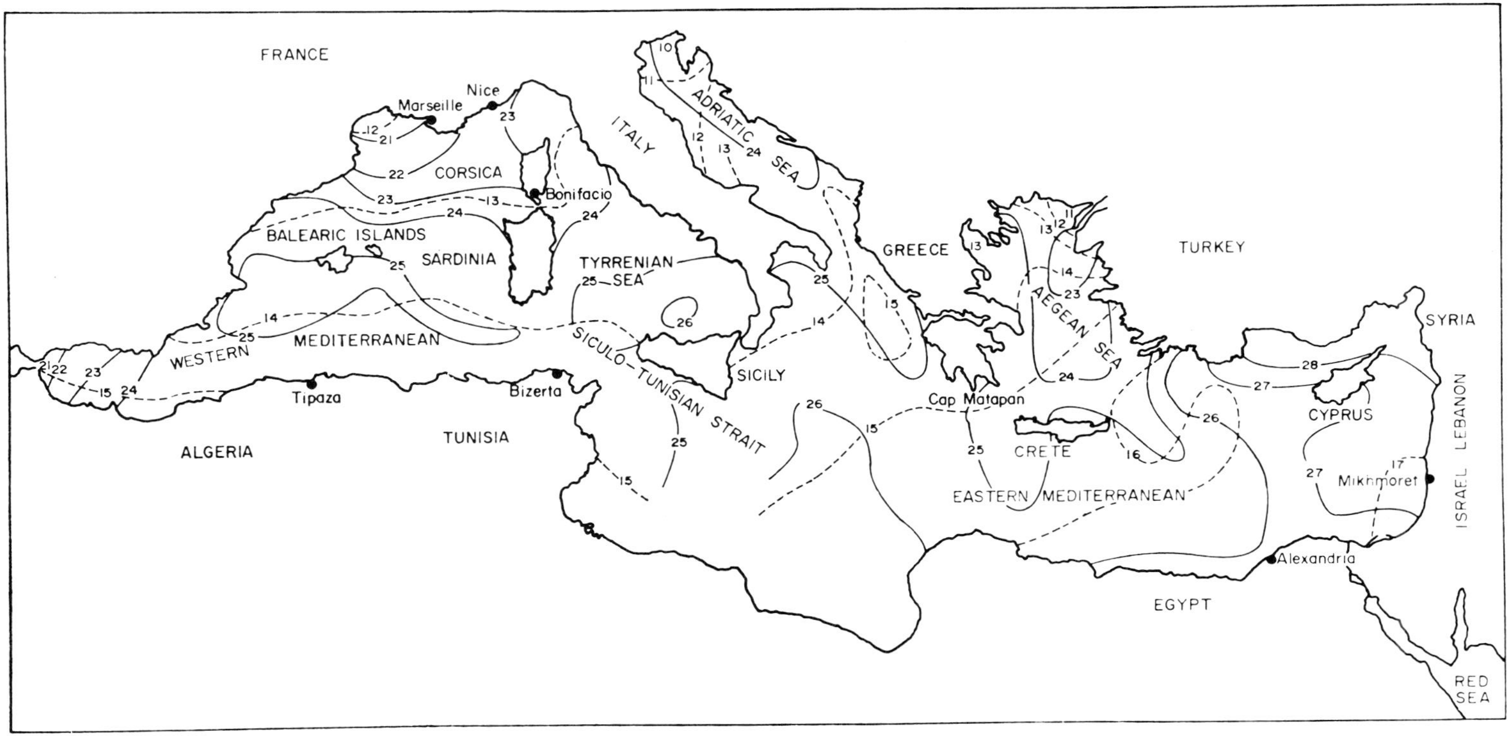

Fig. 29. Mean surface temperatures in August (———) and March (– – – –) in Mediterranean (from Lipkin and Safriel, 1971)

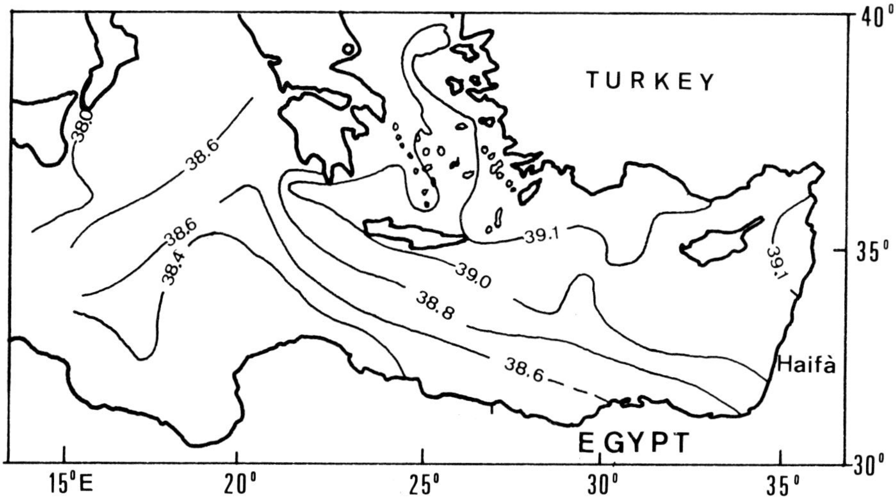

Fig. 30. Winter surface isohalines in Eastern Mediterranean (from Lacombe and Tchernia, 1960)

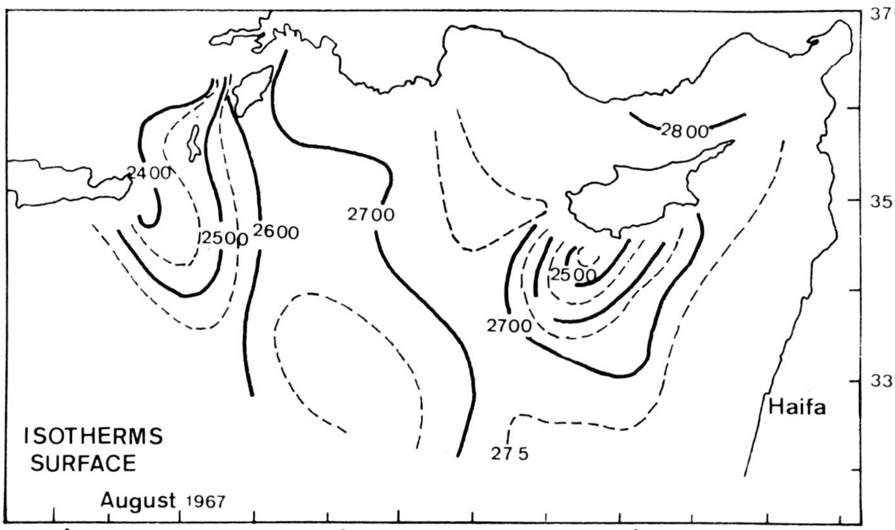

Fig. 31. Surface isotherms in the Levant basin, August 1967 (courtesy Oren)

hydrographic parameters: the annual surface isotherm 20° C and the minimum surface isohaline of 39‰ (39 g/l). These lines are of course fluctuating but they can be considered as an arc which extends from the Nile delta to the eastern tip of Crete and afterwards follows the island arc of the Dodecanese to Rhodes. The shape of this limit is determined chiefly by the horizontal deflection of the isohalines along the Egyptian coast, where the Atlantic current brings relatively low salinity waters into the area. As shown by Lacombe and Tchernia (1960) the

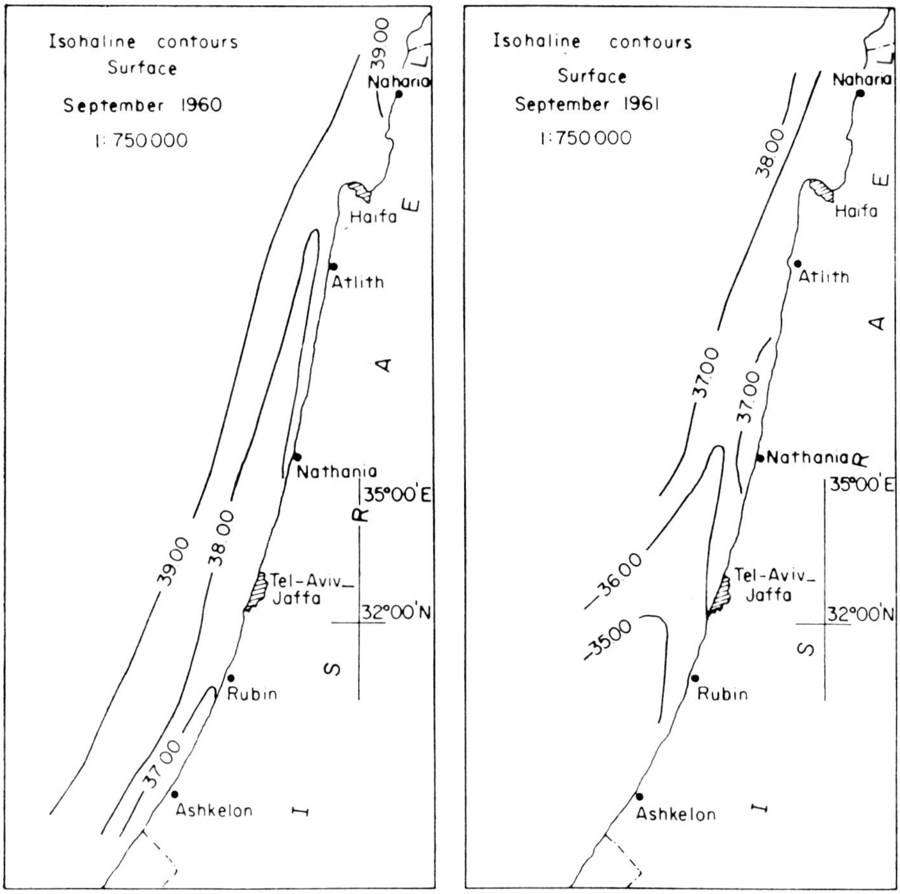

Fig. 32. Isohaline contours off coast of Israel during Nile floods of 1960 and 1961 (from Oren, 1969)

isohaline deflection is maximal in the winter, whereas in the summer the isohalines tend to straighten out in a north–south direction (Fig. 30).

The annual fluctuation of temperatures in the Levant Basin, thus defined, is from a low of 16° C (on rare occasions 15° C in the winter) to a maximum of nearly 29° C (Oren, unpublished data, 1971). The salinity reaches a maximum value of 39.55‰ in the northeastern corner. Within this fairly homogenous basin there are two major hydrographical disturbances: the upwelling of cold waters along the southern coast of Cyprus [summer temperatures of only 22° C (Oren)] (Fig. 31), and the two-month seasonal lowering of the salinity along the north Sinai and Israel coasts because of the Nile floods (Fig. 32) of the past years before the damming of the Nile waters.

Since we are interested in the surface waters to a depth of a few tens of meters, a scheme borrowed from Oren will suffice to show the vertical stratification of the Levantine water masses (Fig. 33). However, the very stable vertical stratification

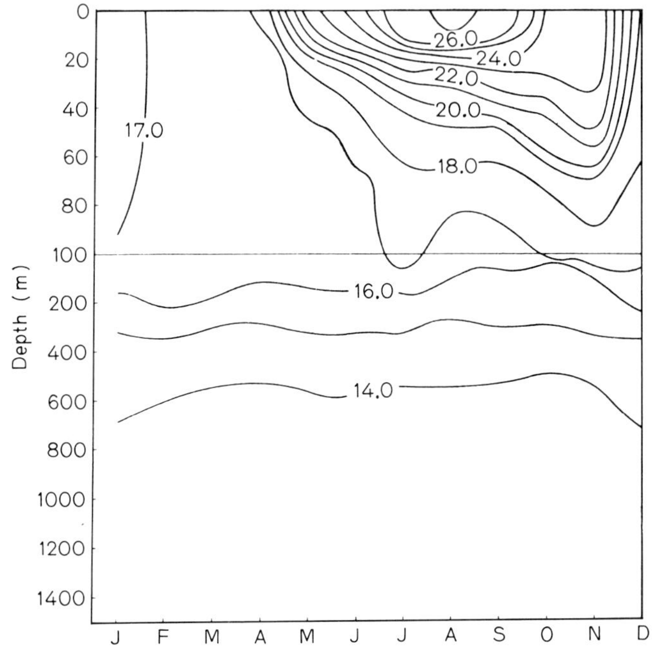

Fig. 33. Seasonal temperature changes in water column of Levant Basin (courtesy Oren)

results, according to Oren, in the impossibility of enrichment of the surface waters by nutrients from the bottom layer.

Of all the marine water bodies of the globe the Mediterranean is known to have the lowest values of nutrient concentrations and of primary production. The mean value of total phosphorus in the Levant basin is about 0.8 µg/l, showing only occasional peaks. The values of nitrate nitrogen along the Israel shore fluctuate between 0 in the winter to 0.6 µg/l during the summer. Near Cyprus, Oren (1966) measured NO_3 values at the surface of less than 1 µg/l. According to Oren these values are comparable to those in the Western Mediterranean.

However, the differences are considerable in the values of primary production. The mean gross yearly production of the coastal waters of Israel is given as 36 mgC/m^2/year (Oren, unpublished data) which according to this author is some 50–60% less than the annual production in the Western Mediterranean. Generally, if values for a certain depth are compared, the values in the Levant Basin are an order of magnitude lower than those of the Western Mediterranean. The Levant Basin fares better if the whole productive water column is compared, since the compensation depth in the Levant is deeper (about 75 m) and the nutrient supply of the deeper water layers is better than in the West. Again, however, following the fate of the Lessepsian migrants in the Levant Basin, we confine ourselves to the upper 60–70 m of the water column.

Thus the Levant Basin is a subtropical, saline "cul de sac" of the warm-temperate Atlanto-Mediterranean zoogeographic area, additionally characterized by the lowest values of primary production recorded in oceanic water. The implications for the animal and vegetal populations are obvious: a clearcut quantitative and qualitative impoverishment with indications of smaller individual sizes.

I could not find comparative data for the planktonic biomasses. Such data are available however for the benthic level-bottom biomasses and they clearly illustrate the poverty of the Levant Basin. Tchukhtchin (1964) found an average benthic biomass of 0.089 g/m^2 in the Eastern Mediterranean between 200–1000 m. For depths less than 200 m he indicates a value of 4.12 g/m^2 on the Sicily–Tunisia sill, as compared to values of 266.4 g/m^2 in the Atlantic, or even 388 g/m^2 in the Black Sea. Gilat (1964) compares the benthic biomass of the circalittoral shelf areas and gives values of 8–16 g/m^2 for the Eastern Mediterranean and 48–126 g/m^2 for the Adriatic. Surprisingly low values are indicated by Tchukhtchin (1964) for Mersin Bay in the northeastern corner of the Levant: 0.008 g/m^2 at 173 m. This is probably a rather extreme case of faunal scarcity, lower than the usual low of the area, and probably based only on macrobenthic biomass. Gilat's (1964) macrobenthos biomass values from the shelf of Israel are in the extremely low range of 0.059 to 0.075 g/m^2. There is only one notable exception in the Levant Basin—or rather at its southwestern limits—the fishery grounds off Alexandria, where Vatova (1935) found 551 g/m^2; Kiseleva and Tchukhtchin (1965) reported 500 g/m^2 for a station near Alexandria. Such high biomasses were probably found all along the shores of the Nile Delta and speak for the local influence of the nutrient rich fluviatile deposits of the Nile. For the Levant Basin as a whole, the nutrient enrichment by the Nile water had always had only a strictly localized and seasonal character.

The impoverishment towards the east is especially well-documented in the qualitative aspect. The species diversity of all the environments is sharply decreasing in something like two steps: the Western Mediterranean→Aegaean Sea→the Levant Basin. There are 118 Atlantic species of the planktonic Hyperiidea (Amphipoda) and 46 species in the Western Mediterranean. In the Eastern basin there are only about 10 species. Eight species of Pteropoda of the Western Mediterranean do not reach the Eastern part. Alvariño (1974), reviewing the Siphonophora, writes: "Many species do not reach the Levant Mediterranean basin, and only appear accidentally there, a few get settled in the area". Recently Godeaux (1974) documented the same paucity with respect to the Thaliacea (salps), mentioning in particular the lack of the genus *Pyrosoma* in the Levant Basin.

For the benthic organisms the impoverishment is much better documented. There are whole categories of habitat-forming organisms which are missing, such as the water plant *Posidonia oceanica* and the wealth of the Western Mediterranean gorgonarians. No single species of Gorgonacea has reached the Levant coast. The impoverishment is equally impressive among the Decapoda and the Echinodermata. Pérès (1958a) analyzes the populations of Ascidiacea along the Israel coast and remarks that the families Molgulidae and Polyclinidae are very scarce or absent. He mentions eight species of Molgulidae in the Western Mediterranean, four species in the Sicily–Tunisia area, one species in the Aegean Sea and none along the Israel coast. The Polyclinidae, which still form 15–20% of the Ascidian diversity in the Aegean Sea, are represented among the 33 species of the Israel coast by only one species, *Amarocium lobatum*.

Tortonese (1951, p. 232) was of the opinion that: "The faunistic poverty of the Eastern Mediterranean is a certain fact not due to the incomplete research ... It is to assume that future research will demonstrate the eastward extension of some species but this will not change the present views on the quantitative relationship between the two great mediterranean basins." The same author (Tortonese, 1973) later emphasizes the fact that since the "Calypso" cruise in the Aegean sea many western species have been discovered there. This only proves the necessity for a good subdivision within the Eastern Mediterranean separating the Aegean from the Levant Sea.

This becomes especially clear in the careful study made by Fredj (1972, 1974) on the distribution of 1244 file-indexed species of benthic invertebrates of the Mediterranean. Unfortunately, when subdividing the Eastern Mediterranean (minus the Adriatic Sea) Fredj distinguishes a North Eastern Mediterranean ("Méditerranée orientale de Nord"), a Central Eastern Mediterranean and a South Eastern Mediterranean. While the northern subdivision is a natural one and includes only the Aegean Sea, the "Central" one extends in a broad zone from the Israel coast near Haifa, along the Anatolian coast, including the southern part of Greece together with the big islands, and ends on the western shore of the peninsula near Corfu. The South-East Mediterranean is only slightly more homogenous since it comprises the shores from Haifa, along Egypt and Libya, ending at Cape Bon which faces the Straits of Sicily.

Even so, Fredj calculates that the Eastern Mediterranean as a whole contains only 48.4% of the 1244 species analyzed while the Western Mediterranean contains 92%. The Eastern Mediterranean has only 4.2% of the species strictly limited to this basin while 35.4% of the species indexed by Fredj can be found only in the Western Mediterranean. Of the 247 endemic species found among the 1244 indexed species (19.8%), 82.2% are found in the Western Mediterranean and only 22.8% in the Eastern. Of the total 247 endemic species, 49.8% are exclusive to the Western Mediterranean and only 4.5% are found exclusively in the Eastern Mediterranean.

With the exclusion of the Aegean Sea and Greek Peninsula, these figures would have been more impressively in disfavor of the species diversity of the Levant Basin. One must also keep in mind the limitation of the method used by Fredj, who artificially selected the animal groups which form his species list: the main groups are the Echinodermata, Polychaeta, Sipuncula, Solenogastres, Polyplacophora, Scaphopoda, Pelecypoda, Cephalopoda, and Decapoda. The Gastropoda are not included, and the Polychaeta, a group in which the distributional information is still incomplete, form more than one quarter of the 1244 species studied. Nevertheless, the zoogeographic study of Fredj (1974) is the first noteworthy attempt to quantify the zoogeographic patterns of the Mediterranean fauna.

Almost all positive traits of the Eastern Mediterranean as a mediterranean subregion result from the phenomenon of Lessepsian immigration, which should be analyzed separately. It is difficult to find any positive trait in the species diversity of the Levant Basin of one disregards the additions which have occurred since the opening of the Suez Canal.

Prior to the opening of the Canal there were practically no species specifically restricted to the Levant Basin. Perhaps the only positive traits can be found in the quantitative increase in the populations of some species typical for warm and saline waters. Parker (1958), for instance, notes the considerable increase in the numbers of the planktonic forminiferan *Globigerinoides ruber* in the Eastern Mediterranean. Here it represents 41–77% of the planktonic Foraminiferida, while in the Western Mediterranean it constitutes only 1–25% of the total. Pérès (1967) emphasizes the exceptionally rich development along the Levant coast of the facies-building mollusc *Vermetus triqueter*, rare in the West. Pérès also emphasizes the rich development of the soft bottom meadows of the green alga *Caulerpa prolifera* accompanied by *Caulerpa scalpelliformis* (considered by Pérès to be a Red Sea immigrant, but known also from the Canary Islands).

A few words must be said about the so-called "Levantine nannism"—the smallness of the specimens in the Levant Basin, especially apparent in groups such as Sipuncula (Stephen, 1958) and Porifera (Levi, 1957). Remarked already by Forbes for the Aegean Sea and by Le Danois (1925) for the East Mediterranean fishes, it is later accepted as a general phenomenon by Tortonese (1951). Such small-sized individuals indicate suboptimal environmental conditions. The smaller biomasses of an impoverished fauna is the general rule for biota of the Levant Basin. Pérès (1967) concludes his study on the Mediterranean benthos with the words: "It appears that certain Mediterranean biocoenoses, above all those in the infralittoral zone, are under a kind of disequilibrium which is a consequence of the non-utilization of a certain fraction of their production"—a statement more applicable specifically to the Levant Basin.

The Lessepsian migrants may be considered, in a figurative sense, "welcome guests" in the impoverished, subtropical cul-de-sac of the Atlanto–Mediterranean world.

3.3 The Levant Shore— the Core Area of the Lessepsian Immigration

Most of the Lessepsian immigrants are to be found along the shores of Sinai, Israel, Lebanon and Syria, thus necessitating a more detailed description of the Levant shelf area, an integral part of the Levant Basin (see Map 2, p. 85).

The Levant shore is dominated by a relatively broad shelf built up by sediments deposited by the Nile. The majority of the 100–130 million tons of silt carried yearly by the Nile (Hammerton, 1972) was deposited to the East of the Delta, by the counter-clockwise Coriolis current. At depths of 30–75 m (Oren, unpublished data) the outrunners of the Atlantic current also contributed to the transport.

The shelf of the Levant shore, from Port Said to the surroundings of Haifa, is relatively wide. According to Gorgy (1966) it reaches a maximum width of 75 km in the Gulf of Pelusium (Gulf of Tina), then narrows to 29 km off the Sirbonic Lagoon (Bardawil), broadens to 40 km at El Arish, and narrows gradually along the Israel shores from 30 km at Rubin to 10 km near Haifa (Emery and Bentor, 1960). Near the Lebanese border, the shelf is very narrow, some 3 km wide, and it remains so along the entire Lebanese coast and southern part of the Syrian coast, broadening somewhat in the Bay of Mersin (Alexandretta, Iskenderun).

The bottom of the shelf is formed from Nile sediments, which predominate in the northeast till near Jaffa; from there, the width of the shelf decreases and the amount of Nile sediments decreases too. The rocky basement layer is covered by the sediment, for most of the shelf to Netanya. Off the Sirbonic Lagoon there is one area of shell bottom with some rocky outcrops between 10 and 50 m depth (Gorgy, 1966). Farther to the north the bottoms again become uniformly sedimentary. Gilat (1964), working north of Ashdod, did not encounter rocky bottoms. Around Netanya coralligenous bottoms start to appear regularly at depths of about 20 m. Gilat (1964) suggests that some "islands" of coralligenous bottom might occur even south of Netanya, at depths of 80–90 m. The isobaths as well as the shore line are remarkably straight along the coast of Israel. Only the Bay of Haifa, protected by the Carmel ridge and its calcareous promontory has a more complicated bottom structure with an alternance of coralligenous reefs between 10–20 m, partly formed by the coral *Dendrophyllia cornigera*, and of level bottoms covered with meadows of *Caulerpa scalpelliformis*. The shallow 4–5 m deep bottoms in the old port of Acre are covered by pearl oyster beds (*Pinctada radiata*). North of Haifa Bay the shelf becomes steep and predominantly rocky.

According to Gilat (1964), the sequence of the sedimentary bottoms north of Ashdod is as follows: the bottom is sandy to approximately 20 m; the sediment is sand mixed with equal amounts of silt and clay around 35 m; at 46 m the bottom is predominantly silt and clay. The organic content, plant detritus deposited by the Nile, is relatively high on the southern shores of Israel i.e., up to 0.86% of the sediment (Gilat, 1964). Shell bottoms start to occur off Netanya.

The shore is almost uniformly sandy till Mikhmoret, a point north of Netanya (Safriel and Lipkin, 1975). Only one rocky outcrop is found along this long 400-km stretch from Port Said to Mikhmoret: the rocky promontory of Yavne Yam (Palmahim) south of Tel Aviv. From the area of Ashkelon northwards slabs of beachrock appear, obliquely covering the bottom immediately below the sea level.

The beachrocks, wave-battered and permanently corroded by sand, can harbor only a limited number of animals. Typically rocky littoral environments appear only from Mikhmoret northward. However, as soon as rocky ridges of sandstone appear, the ridges are accompanied by organogenic rock platforms of the vermetid molluscs, *Vermetus triqueter* and *Dendropoma petraeum* (Safriel, 1975).

The Levant shore is practically tideless, differences between high and low tides usually not exceeding an amplitude of 30 cm. At high air pressure and eastern winds the level of the sea is exceptionally lowered by several tens of centimeters.

The current is generally directed counterclockwise and is relatively weak. In the Port Said-El Arish area, Gorgy (1966) indicates for October 1959 and 1960 a velocity of 4.5 knots near Port Said, which slows down through a series of anticyclonic branches near the Sirbonis to only 0.8–1.0 knots along the shore at El Arish (Fig. 34). This current is further slowed down along the Israel coast: Liebman (1935) gives a maximum value 0.6 knots. Oren and Komarovsky (1961) mention a velocity of 0.20 knots (0.1 m/sec) as most frequent along the Israel shores. The

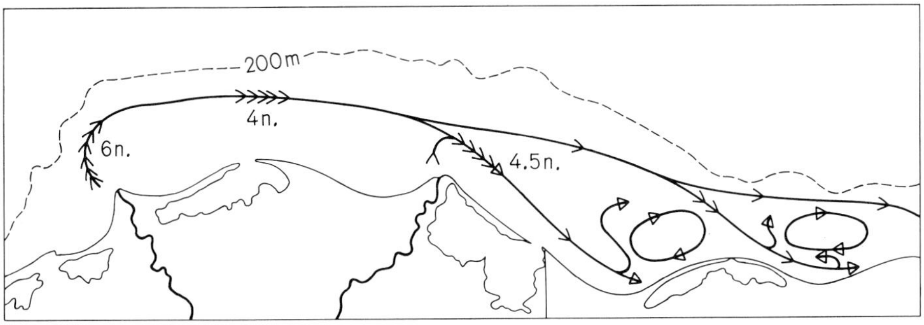

Fig. 34. Surface current patterns and speeds (in knots), off Nile Delta and Port Said (from Gorgy, 1966)

values found by Hecht in 1964 (pers. comm.) were in the range of 0.30 knots. Toward the coasts of Lebanon and Syria the speed of the current increases and reaches 1 knot (Gruvel, 1931). Oren (1970) concludes that along the coasts of Israel the currents are weak, and although there is a permanent longshore current induced by Coriolis force, the direction of the currents is very much influenced by winds.

Under such circumstances, as first calculated by Liebman (1935), it took six weeks for the Nile waters to reach the Central coast of Israel. The dams on the Nile mouths were opened in the first days of August, but the outpouring waters reached the coast of Israel only in mid-September.

If we take into account that a passively carried planktonic larva leaving the area of Port Said spends some six weeks reaching the first extensive rocky shores near Mikhmoret, only animals which could settle on the level bottoms or the beachrock on the way could successfully overcome this obstacle.

The temperatures along the Levant shores are probably the major factor in determining the success or failure of the Lessepsian immigrants in their new environment. The low winter temperatures are a serious limiting factor for the build-up of the immigrant populations. Liebman (1935) followed the annual temperatures near Haifa during 1929–1930 and found a minimum of 15.4° C in January and a minimum monthly mean of 16.6° C in February. Oren gives temperatures continuously measured at Haifa since 1947. He considers the months of February and March as the minimum temperature months. After the warm winters of 1954–55 and 1955–56, winter temperature did not descend below 17° C. This change of about 2° C in the winter minima is correlated by Oren (1957) to a "sudden increase" of certain Indo-Pacific immigrants. According to Oren (1970) the average inshore minimum over the last 20 years was 17.4° C. However, a temperature of 14° C was measured in 1959 near Ashdod.

Minimum temperature off Beirut between 1936–1939 was 17.1° C according to Rouch (1945). The summer maximum temperature there exceeded values of 28° C and Liebman (1935) reported 29.4° C for the month of August. (All these values are surface temperatures and do not change radically in the upper 30 m layer, which is of significance to our subject.)

Much has been written about the diluting influence of the Nile floods. Some 85% of the 84 billion m^3 yearly discharge reached the sea in the months August–

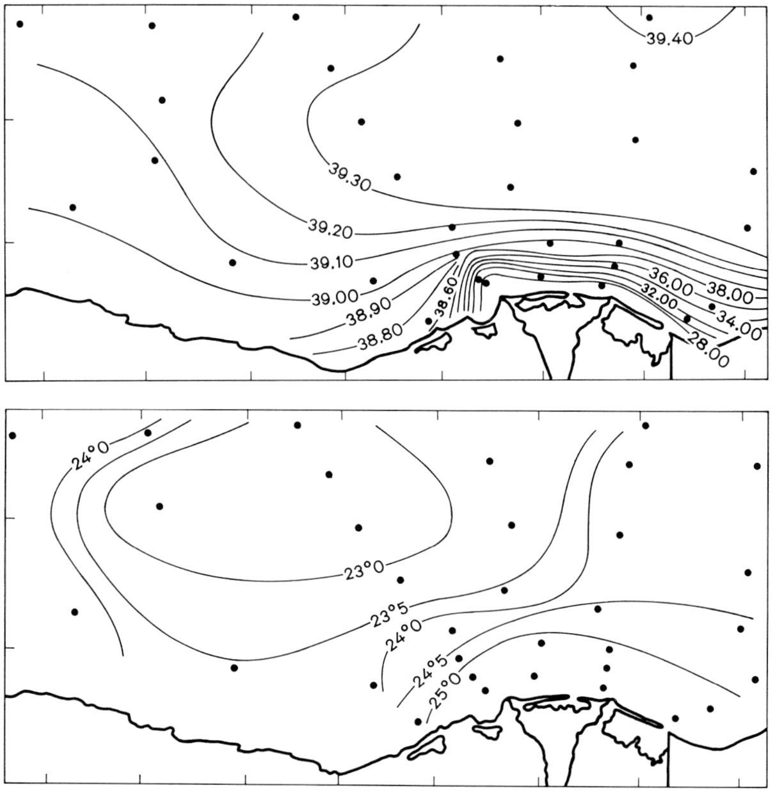

Fig. 35. Autumn salinities and temperatures off Egyptian coast at 25 m depth (from Gorgy, 1966)

November. Such an annual discharge is documented as far back as 600 A.D., but existed even in classical Egyptian times. Several hundred years ago the two active branches, Rosetta and Damietta, were dammed during the low water season by the earthen dams of Edfina and Faraskur. In August they were opened and the waters gushed out into the Mediterranean. The two branches were opened at an interval of several weeks. The Damietta branch, 60 km from Port Said, recently carried only half the water carried by the Rosetta branch which opens about 200 km west of Port Said.

The outpour of the Nile water through the two branches was felt in the form of two salinity lows, clearly distinguished even along the coast of Israel (Oren, 1969).

Soon after the opening, the surface near Port Said reached salinities of 30–35‰, sometimes even 29‰ (Gorgy, 1966; Morcos, 1967a) (Fig. 35). Oren (1969) reports one instance when 25‰ was measured off Gaza in 1947.

Liebman (1935) measured a drop of salinity off Haifa to 34.02‰ in September 1930. Rouch (1940) measured 35.4‰ in 1938 off Beirut. More recently the minimum values were higher. For September 1961, Oren (1969) indicates 35‰ as far north as Tel-Aviv. The tongue of low salinity water normally did not extend beyond the southern border area of Lebanon. By the end of October, i.e. after two months, the salinity was back to normal, over 39‰ (Fig. 32).

The influence of the Nile flood was also felt in its biological consequences. Gorgy (1966) measured a very high value of PO_4 phosphate: 1.76 μg/l at 30 m depth in October 1959 in the Gulf of Pelusium (Tina). This was accompanied by a water bloom of planktonic algae reaching a maximum of 3,120,000 cells/l. In March 1959 the phosphate levels were low (maximum 0.11 μg/l) and the maximum photoplankton counts were only 8100 cells/l.

Phytoplankton blooms accompanying the Nile waters were also observed by Liebman (1935) off Haifa. Komarovsky (1953), Halim (1960), and Halim et al. (1967) listed the diatom species responsible for the Nile blooms, and the first author indicated the cladoceran *Podon polyphemoides* as typical of the September plankton. The main pelagic commercial fish of the Levant Basin, *Sardinella aurita* concentrated preferentially in the offshore area of the Nile Delta during the flood and bloom period.

There is no evidence of negative influence of the Nile flood on the Lessepsian migration. After all, the species advancing northward through the canal had already encountered low salinities of ±38‰ in Lake Timsah, and the low values to the east of Port Said were restricted to two months, September–October at most, and concerned only the topmost 5 m of water column. West of Port Said, where the Nile flood could dilute the inshore areas till February–March, the "low salinity" barrier was most probably a more effective one.

The Nile was hermetically dammed behind the Aswan High Dam in 1965. All the above-mentioned seasonal phenomena have ceased since, and since that date hardly a drop of Nile water has reached the Mediterranean. For reasons explained above no short-term influence on the eastward migration of the Lessepsian migrants is expected. However, the westward migration along the African shore may be facilitated, as shown previously (Por, 1971b).

In the long run, the shoreline of the southern Levant coast is expected to be gradually eroded, since the sediment supply has been cut off. The Sirbonic Lagoon may become an open gulf in the not too far future. However, even before that, a slow increase in the salinity of the Levant Basin will presumably follow. With an annual evaporation of 1300–1500 cm/year the loss of the Nile waters will start to be felt. A more saline Levant Basin will eventually be more favorable to the Lessepsian migrants—provided temperatures do not decrease. Oren (1969) expects a slight increase in the primary production due to the disappearing seasonal saline stratification of the Levant inshore waters. However, these forecasts will be discussed in further detail in the concluding Chapters 3.7 and 3.11.

3.4 The Analysis of Lessepsian Migration—Methodology

The two lists of Lessepsian migrants contain a total of 204 species[1], of which 128 are considered as High Probability Lessepsian Migrants (HPLM) and 76 as Low Probability Lessepsian Migrants (LPLM). The few cases of parasites reported till now as having migrated together with their hosts are not included in these lists.

---

[1] New additions while this manuscript was in preparation are the fishes *Rhonciscus stridens* and *Gobius ? lesueuri* (Ben Tuvia personal communication).

For a statistical analysis the number of migrant species is too small, since specific information on each species of this extremely diversified list is quite unequal. Therefore, number symbols have been used in an attempt to express the parameters of the time-distance analysis which will be presented below. The analysis of the faunal diversity of the migrants and of their success as colonizers will, for purposes of clarity, follow the time-distance analysis, rather than precede it.

The information compiled here is by necessity based on incomplete records, an uneven amount of data, both in time and space with regard to the different taxa, as is the raw material on which every zoogeographical analysis of whole faunas is always based. There has never been a sustained effort in this area criss-crossed by political and cultural boundaries. The different taxa of Lessepsian migrants, besides the fact that they behave differently on the zoogeographical scene are also better or less known on the taxonomic and distributional level.

Nevertheless, an attempt to quantify the conclusions has been made, not only to avoid wordiness or to better express conclusions which were known a priori, but also in the hope of obtaining new insights into the migration phenomenon.

3.5 Time–Distance Analysis

193 of the 204 species were divided into the categories reported below (see Table 5). Of these, 119 or 61.65% are known to have progressed through the Suez Canal before invading the Mediterranean. The remaining 74 species or 38.35% have not been reported from the Suez Canal either due to incomplete records or because they passed the Canal in "one jump" (without having established stable populations in the waterway). Since it is reasonable to assume that at least half of these 74 species do in fact have (or did have) Suez Canal populations, one can assume that the great majority of the Lessepsian migrants (tentatively 80%) are indeed recruited from species which were able to build up populations in the Suez Canal environments.

A second point to be analyzed is more complicated: namely, the correlation which exists between three parameters: (1) the date of settlement of a given species in the Suez Canal; (2) the speed of its migration into the Mediterranean; and (3) the area it occupies in the Mediterranean.

For these three parameters I have used a set of three-digit symbols, ranging in the first two cases from "early" to "late", and in the third case from "close" to "distant". The species not reported from the Suez Canal are designated by 4.0.0. In a 3-digit symbol each of the three parameters will have their specific position.

1. The first parameter, "date of the first report in the canal", is subdivided as follows:

1.0.0 Species reported from the canal before 1906. This period includes the collections by Fuchs, Keller, Krukenberg, Tillier and Bavay (see Chap. 2.2).

2.0.0 Species reported from the Canal before 1930. This period includes the extensive collections of the Cambridge Expedition of 1924.

3.0.0 Species reported in the Canal after 1930, including mainly the species of Gruvel and Moazzo, Lami, and the Hebrew University-Smithsonian Institution collections.

4.0.0 Species not reported from the Suez Canal.

2. The second parameter, "emergence into the Mediterranean", is graded as follows:

0.1.0 Species reported in the Mediterranean before 1929. This includes the Cambridge Expedition reports, and the reports by Pallary, Gruvel, and W. Steinitz.

0.2.0 Species reported in the Mediterranean before 1955. This is the year which according to Oren (1957) marked the beginning of an upward fluctuation in the temperature of the Levant Basin.

0.3.0 Species reported in the Mediterranean after 1955.

3. The area occupied in the Mediterranean is graded into four categories, according to a subdivision into provinces based on preliminary information:

0.0.1 Species confined to the area off Alexandria in the west and the Sinai coast in the east. These are species which did not leave or could not settle beyond the immediate vicinity of the Canal outlet.

0.0.2 Species which have advanced along the Levant coast and have been reported along the shores of Israel, Lebanon, and Syria.

0.0.3 Species which have reached the area of Iskenderun and Mersin in the eastern corner of Anatolia, and have advanced as far as Rhodes, or have been reported from the islands of Cyprus and Crete.

0.0.4 Finally, a somewhat odd category—not a further gradation of the previous three—are the species which are known from the North African coast, west of Alexandria, or from the Islands of Sicily, Malta, Lampedusa, etc., i.e. the western confines of the Eastern Mediterranean.

Taking into account both the Suez Canal species and those not reported from the Canal, there are 48 possible combinations. These are presented in a "checkerboard" diagram (Table 5) which also indicates the respective number of species. The 3-digit symbol designating each species is also included in the two lists of migrants (see Tables 2 and 3).

From the list of 204 migrants, only 193 species have been provided with such symbols. The remaining 12 species have not been included in this analysis for different obvious reasons. These are species which still have an uncertain status, or represent a group exceptional from the ecological point of view.

A last point worth mentioning is that in the strictly chronological sense, the 2.0.0 species are contemporaneous with the 0.1.0 species and the 3.0.0 species with the 0.2.0 and 0.3.0 species. The fact that only 1 species out of 119, the bivalve *Chama broderipi* is a 3.1.0 species, i.e. a species reported from the Mediterranean *before* being reported from the Suez Canal, is encouraging for the reliability of our knowledge. In addition *Chama* is a 3.1.1 species, i.e. one which did not expand beyond the immediate vicinity of the Canal outlet.

If we apply the same to the 4.0.0 species, we shall find only five 4.1.0 species (vs. 31 4.2.0 and 38 4.3.0). Of these, two are algae: *Caulerpa scalpelliformis* which was reported from the Mediterranean in 1930 (at the limit of 0.2.0!) and *Hypnaea nidifica*, a species probably passively transported; the sponge *Mycale erythaeana*

Table 5. Time-distance analysis of 193 species of Lessepsian migrants

Category	No. of species	Category	No. of species	Category	No. of species	Category	No. of species	Category	Total No. of species
1.1.1	1	2.1.1	1	3.1.1	1	4.1.1	3		
1.1.2	2	2.1.2	5	3.1.2	0	4.1.2	2		
1.1.3	1	2.1.3	2	3.1.3	0	4.1.3	0		
1.1.4	5	2.1.4	0	3.1.4	0	4.1.4	0	0.1.0	**23**
1.2.1	1	2.2.1	11	3.2.1	0	4.2.1	10		
1.2.2	12	2.2.2	9	3.2.2	4	4.2.2	13		
1.2.3	3	2.2.3	5	3.2.3	1	4.2.3	2		
1.2.4	1	2.2.4	1	3.2.4	2	4.2.4	6	0.2.0	**81**
1.3.1	3	2.3.1	5	3.3.1	3	4.3.1	2		
1.3.2	7	2.3.2	15	3.3.2	11	4.3.2	25		
1.3.3	2	2.3.3	1	3.3.3	1	4.3.3	8		
1.3.4	1	2.3.4	0	3.3.4	2	4.3.4	3	0.3.0	**89**
1.0.0	**39**	2.0.0	**55**	3.0.0	**25**	4.0.0	**74**	0.0.1	**41**
								0.0.2	**106**
Known from the Suez Canal					**119**	Not known from the Suez Canal	**74**	0.0.3	**25**
								0.0.4	**21**

Key to symbols:

1.0.0 Reported from the Suez Canal before 1906
2.0.0 Reported from the Suez Canal before 1930
3.0.0 Reported from the Suez Canal after 1930
4.0.0 Species not reported from the Suez Canal
0.1.0 Reported from the Mediterranean before 1929
0.2.0 Reported from the Mediterranean before 1955
0.3.0 Reported from the Mediterranean after 1955
0.0.1 Species found only from Alexandria to Sinai coast included
0.0.2 Species found along the Levant coast
0.0.3 Species found also along the Anatolian coast, Cyprus, Rhodes and Crete
0.0.4 Species found West of the Alexandria–Peloponnesos line

and the bryozoan *Membranipora savartii* both passively transported and also restricted migrants (0.0.1); the last of the five is the swimming crab *Charybdis helleri*. Thus for the great majority of the Lessepsian migrants the rule is prior presence in the Suez Canal, before entering the Mediterranean.

What is the number of future Lessepsian migrants in each of the three periods which we have noted? In the first period (prior to 1906) 39 of the species found in the Canal (1.0.0 species) ultimately became Lessepsian migrants. In the second period (1906–1930) 55 species (2.0.0) were added. The increase in the number of migrants found in the Canal during the third period (after 1930) was only 25 species (3.0.0). These results can be interpreted in two complementary ways:

1. There was an increase in the number of potential migrants within the Canal fauna, while there were not enough new entries of potential migrants from the

Red Sea to counterbalance the number of species which had migrated into the Mediterranean. A simpler explanation is that there is a limited stock of Suez Canal species from which the migrants are recruited; the biota of the Canal has reached a climax-like stability and, therefore, the migration as far as species diversity is concerned, is approaching an end. This is a thesis already expressed by me in previous papers (Por, 1971 b, 1973 c) and still upheld here.

2. The chronological difference between 3.0.0 and 0.3.0 is much shorter than between the chronologically earlier combinations. Eventually, *only* the 3.0.0 species managed to emerge from the Canal during the last decades.

There is no evidence for a presently increased rate of migration *into* the Mediterranean as presumed by many authors. In the two first periods (0.1.0 and 0.2.0) the increase is evident: 23 species as compared to 81, respectively. Between the second the third period (0.2.0 and 0.3.0) the increase is much less dramatic, i.e. from 81 to 89 species. Combining this with the circumstance of considerably less "new entries" into the Canal (i.e. 25 3.0.0 species) the impression of a general slowing down of the process of Lessepsian migration becomes more apparent.

Nevertheless, it must be noted that the difference between 0.2.0 and 0.3.0 is mainly due to the large number of 4.3.0 species, i.e. not reported from the Canal. If one takes into account that the species which could cross the Canal quickly, in "one jump", are among the 4.0.0 species, a tentative conclusion can be presented that in the last years the "one jump" migration might have become somewhat more important compared to the "step by step" process of migration. In this connection it is of interest that among the latest additions to the Lessepsian migrant fishes are *Rastrelliger kanagurta* and *Scomberomorus commerson* (two 4.3.2 species)—the first cases of outright pelagic fishes (sensu Ben-Tuvia, 1972) which have crossed the Canal. Also the two pelagic diatoms, reported quite recently by Kimor (personal communication), *Chaetoceros coarctatus* and *Rhizosolenia indica*, are the first instances of Lessepsian migrants among phytoplanktonic organisms.

With regard to the localization of the Lessepsian migrants in the Mediterranean, several other facts become evident (Fig. 36). There are only 41 species out of 193 (21.24%) which are 0.0.1 species, i.e. confined to the vicinity of the Canal outlet. It is clear that if the Egyptian coast is to be considered a "take-off area" for the migrants, they obviously do not spend too much time there before moving farther. One also has to take into account that this area is not too favorable as far as substrate conditions are concerned. Some of the 0.0.1 species are euryhaline, for which the Sirbonic lagoon or Lake Manzala are favorable environments. Another group among these is that of the 10 epiphytic species of algae reported by Aleem from the Egyptian and Sinai coasts. Some of the 0.0.1 species have been found during the extensive benthic research by Steuer (1935) on the fishery grounds of Alexandria.

The great majority of the Lessepsian migrants are found along the Levant shores (0.0.2)—106 species have reached the Levant shore (54.92%). Among the 0.0.3 species are 31 other species which can be found on the Levant shore. The 0.0.4 area of North African shores has four species in common with 0.0.2. This brings the number of Lessepsian migrants living along the shores of Israel–Lebanon–Syria to 141, or 71.21% of the total.

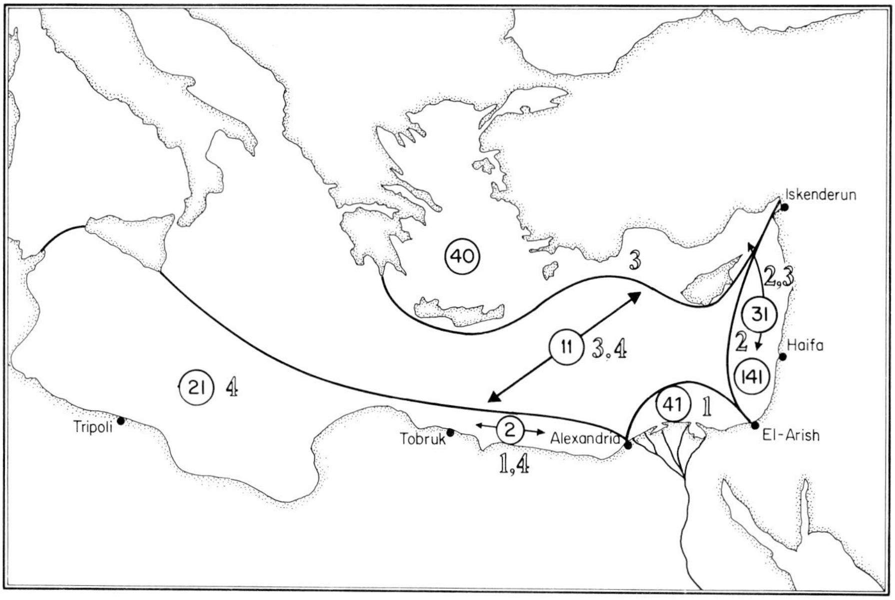

Fig. 36. Distribution of the Lessepsian migrant species in the Eastern Mediterranean. *Numbers:* indicate the four areas (see Table 5); *numbers in circles:* the number of species (original)

Only 40 species have reached the Anatolian coast, Rhodes, Cyprus, and the Crete-Santorin area. The statement found in several papers that Red Sea species (i.e. Lessepsian migrants) have spread into the Aegean Sea is generally wrong. The farthest point reached along the Anatolian coast is the Island of Rhodes. From the area of Izmir, Artüz recorded three fishes (according to Ben-Tuvia, 1972): *Saurida undosquamis*, *Siganus luridus*, and *Upeneus moluccensis*. Another fish, *Sphaeroides spadiceus* was reported in 1952 from the Island of Samos by Anandiades. In fact, in the zoogeographical sense the Aegean Sea is defined by Pérès (1967) and by Fredj (1972) only as the sea *north* of the line connecting the Islands of Euboea and of Andipsara; Izmir is approximately *on* this line. There are other geographical terms used to describe the sea area between Rhodes and Crete and the Greek mainland south of the Andipsara line, e.g. the Sea of the Cikladhes or the Sea of Crete. This is still an area of warm-water fauna, according to Pérès (1967); the cold-water fauna characteristic of the Aegean Sea starts only north of the Andipsara line. Beyond it there are no reports to date of Lessepsian migrants.

Of the 29 0.0.3 species and an additional 11 species found both in the 0.0.3 and 0.0.4 areas (i.e. from the 40 species which have reached the Anatolian coasts and the Islands), 31 are found also along the Levant coast (0.0.2 area). A gradual spread of the Lessepsian migrants may thus be postulated: first northward and then westward along the Levantine and Anatolian coasts. Of the nine species which are not shared with the 0.0.2 area, *Halophila stipulacea* is probably a Pre-Lessepsian settler; the crab *Ixa monodi* is known only from its type locality—Mersin Bay; and the snail *Nerita sanguinolenta* from Karpathos is a rather new addition, due to a taxonomic opinion by Mienis (1974). The rest are polychaete species where, on one hand, there may still be taxonomic revisions and, on the other hand, the distribution is not well known. For example, one of these worms, *Tharyx dorsobranchialis* was recently reported from the Mediterranean coast of Sinai by Ben-Eliahu (1972d).

The gradual longshore advance of the Lessepsian migrants along the Asiatic coast is well-documented. This can be seen if we compare the Lessepsian species of Cyprus with those of Rhodes. Cyprus is much nearer to the outlet of the Suez Canal at Port Said, but much more distant from the mainland, while Rhodes is much farther from the Canal, but nearer to the mainland. There are 18 species known from Cyprus and 11 species from Rhodes. The cold waters off the Cyprus coast probably play a role (Gilat, 1974) in diminishing the Lessepsian fauna of this island which is situated offshore from the Levant coast with its wealth of Lessepsian migrants.

The 21 0.0.4 species are a somewhat odd mixture; 11 of them are also found along the Anatolian coast (0.0.3). The fishes *Leiognathus klunzingeri*, *Pranesus pinguis*, *Siganus luridus*, *S. rivulatus*, *Holocentrus ruber*, *Sphyraena chrysotaenia*, *Stephanolepis diaspros*, and *Upeneus moluccensis*, are among the most successful migrants. *Caulerpa racemosa* and *Pinctada radiata* are possibly Pre-Lessepsian settlers, the report of the cowrie *Erronea caurica* is of doubtful precision. Doubtful are also some of the exclusively 0.0.4 species, such as the molluscs *Lophiotoma indica*, *Quoyula madreporarum*, *Umbonium* cf. *vestiarium*. The ascidian *Metrocarpa nigrum* is in Pérès' (1967) opinion probably a Pre-Lessepsian settler in the Mediterranean. That leaves three algae and the report of the fish *Abudefduf saxatilis* from the Bay of Naples. It seems, therefore, that only the eight species of fishes are reliable migrants to the West, as they are found all over the areas of the Eastern Mediterranean which are influenced by Lessepsian migration. Needless to say, there is no clearcut case of a species which went *only* westward along the African coast.

The most successful 25 migrants in terms of area extension, from among the High Probability Lessepsian Migrants (HPLM), are listed below:

Plants:

Hypnaea valentiae
Halophila stipulacea

Decapoda:

Charybdis longicollis
Metapenaeus monoceros
Myra fugax
Penaeus japonicus
Penaeus semisulcatus
Portunus pelagicus
Thalamita poissoni
Trachypenaeus curvirostris

Mollusca:

Brachidontes variabilis
Malleus regula

Pisces:

Dussumiera acuta
Hemiramphus far
Holocentrus ruber
Himantura uarnak
Leiognathus klunzingeri
Pranesus pinguis
Saurida undosquamis
Siganus luridus
Siganus rivulatus
Sphaeroides spadiceus
Sphyraena chrysotaenia
Stephanolepis diaspros
Upeneus moluccensis

A possible addition:

Pinctada radiata (Mollusca)

The success of these species can be seen also in our case as a result of the time factor. If we calculate the percentage of the 0.0.3 and 0.0.4 species—the most distantly spread Lessepsian migrants—as a function of their early or late migrational advance, the results are convincing: from among the species which entered the Canal in the first period (1.0.0) 33.33% are 0.0.3 and 0.0.4, as compared with 16.36% in the second period (2.0.0), and 20% in the third period (0.0.3). The situation is similar when we check the percentage of the 0.0.3 and 0.0.4 species in relation to the period of their appearance in the Mediterranean: they are 34.78% among the 0.1.0 species, only 25.92% among the 0.2.0, and 21.34% among the 0.3.0 species. This too indicates that the spreading of the Lessepsian migrants is a gradual migrational process in which passive or adventive transport does not play a significant role; the relatively high euryhalinity of the early settlers of the Canal might have been beneficial—but at any rate not detrimental—to their further move into the Mediterranean.

3.6 The Diversity of Lessepsian Migrants

After more than a century of Lessepsian migration, some qualitative aspects can already be discerned, in addition to the merely quantitative time–distance aspects.

The great majority of the migrant species belong to a very small number of major taxa (Table 6). If we distribute the High Probability Lessepsian Migrants (HPLM species) according to the taxa then from the 128 species, the Crustacea represent 27.34%, Mollusca 21.09%, and Pisces 21.09%; the plants, from planktonic algae to the flowering plant *Halophila* represent 10.93%, the rest is divided between only six other phyla (Porifera, Coelenterata, Polychaeta, Bryozoa,

Table 6. Relative importance (%) of the eight main taxa among the migrants

Taxon	HPLM (128 sp.)				LPLM (76 sp.)				HPLM + LPLM (204 sp.)	
	No. of species		%		No. of species		%		Total %	
Plants	14		10.93		11		14.47		12.25	
Porifera	5		3.90		2		2.63		3.43	
Polychaeta	9		7.03		14		18.42		11.27	
Crustacea (Total)	35		27.34		16		21.05		25.00	
Decapoda		23		17.96		8		10.52		15.19
Mollusca (Total)	27		21.09		17		22.36		21.56	
Pelecypoda		13		10.15		6		7.89		9.31
Gastropoda		13		10.15		11		14.47		11.76
Bryozoa	4		2.89		3		3.94		3.43	
Ascidiacea	4		2.89		3		3.94		3.43	
Pisces	27		21.09		5		6.57		15.68	
	125		97.06		71		93.38		96.00	

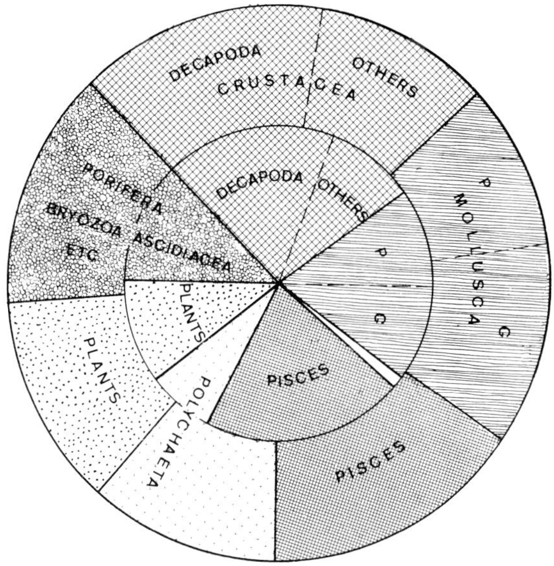

Fig. 37. Circle diagram, indicating proportions of the different taxa out of the total of Lessepsian migrants. *Inner circle:* high probability migrants; *outer circle:* low probability migrants. *P*: Pelecypoda; *G*: Gastropoda (original)

Chordata, Echinodermata). On the class level there are four main classes among the migrants: Pisces Teleostei, 27 species (21.09%); Crustacea Decapoda, 23 species (17.96%); Mollusca Gastropoda, 13 species (10.15%); Mollusca Pelecypoda, also 13 species (10.15%).

If the Low Probability Lessepsian Migrants are added, the total is 204 species and the percentages per phyla are somewhat modified in favor of the Algae and the Annelida Polychaeta: Crustacea 25.00%; Mollusca 21.56%; Pisces 15.68%; Algae 12.25%; and Annelida 11.27%. The total of the five principal phyla becomes 85.76% as compared with 87.48% (their relative frequency among HPLM alone) (Fig. 37).

This indicates that among the algae and polychaetes there are relatively more cases of doubtful migrants than among the other phyla, especially among the well-studied fishes. However, the relative importance of the five major groups remains unchanged.

Some missing phyla or scarcely represented phyla among the migrants need special mention. First, several planktonic classes and phyla are absent altogether. Echinodermata are represented by three species only and Coelenterata by two or three doubtful species. This diversity pattern parallels the picture found in the biota of the Suez Canal. Without going into a detailed comparison of the Canal biota with the Lessepsian migrants it is clear that every major taxon found among the migrants is also found in the Canal.

The inequality or uneven distribution between the major taxa, as represented within the Canal biota, is even more pronounced at the level of the Lessepsian migrants. Echinoderms and coelenterates are represented in the Canal by several

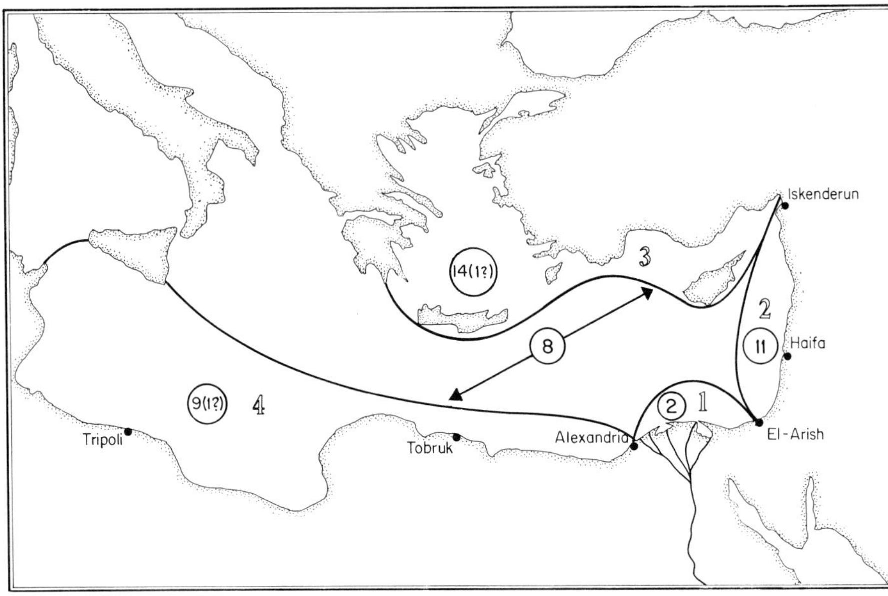

Fig. 38. Distribution of migrant fishes in Eastern Mediterranean (legend as in Fig. 36) (original)

more species of Red Sea origins, and even representatives of the planktonic Chaetognatha are present in the Canal. Red Sea Cephalopoda and Scaphopoda have been reported in the Canal but not further north.

If we look at the composition of the 23 species which migrated into the Mediterranean before 1929 (0.1.0 species) we find only one decapod, the crab *Portunus pelagicus*, while the other four phyla are much better represented. During the following period, after 1930 (0.2.0), no less than 18 species of Decapoda emerged from the Canal; therefore, among the 0.2.0 species the Decapoda alone represent 22.2% giving perhaps an indication that this group actually gained its preponderant role only in the second stage of the Lessepsian migration.

It is interesting to see how the five major groups (Algae, Polychaeta, Decapoda, Mollusca, and Pisces) behave, as far as the area extension in the Mediterranean is concerned. Understandably the fishes are the Lessepsian migrants which have spread farthest to the west, to the Gulf of Taranto, Lampedusa and Tunisia (Fig. 38). As a group, the Decapoda are the second best colonizers, especially the Penaeidae with four species, the Portunidae with three species and *Myra fugax* of the Leucosidae. Decapods have as yet barely been reported from the Western area (0.0.4); they have not advanced along the Anatolian coast beyond Rhodes. *Portunus* might have reached Sicily adventively (Fig. 39).

The Mollusca occupy the third place as successful colonizers: only *Brachidontes variabilis* reached Sicily (but did not advance along the Anatolian coast) and *Malleus regula* has been reported from Cyprus. The bulk of the molluscs is still confined to the Levant shores (0.0.2). There are however some rather dubious

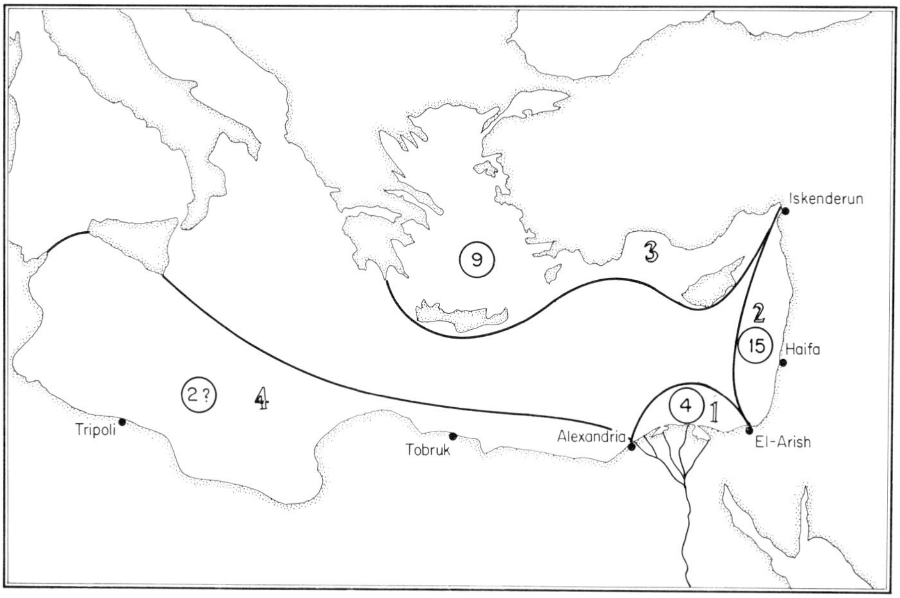

Fig. 39. Distribution of migrant decapod crustaceans in Eastern Mediterranean (legend as in Fig. 36) (original)

cases of westward migration among the molluscs which need further confirmation. *Pinctada radiata* is a special case (Fig. 40).

The algae and polychaetes behave in a way which perhaps might be indicative of possibilities of passive transport. The Lessepsian migrants of these taxa appear without any obvious justification of active distribution in isolated and sometimes distant spots. The best example are the species of eventual migrant polychaetes reported by Harmelin (1969) from Crete, their only known locality in the Mediterranean. Similar is the case of the alga *Hypnaea valentiae* reported only once by Reinbold (1898) from Rhodes, or the case of the alga *Gracilaria arcuata* reported by Feldmann (1931) from Tunis (but also from Alexandria by Aleem, 1948!).

With the exception of these two taxa, it seems that the Lessepsian migrant species advance gradually along the shores. Even in such groups as Bryozoa and Ascidiacea, where passive spread by ships can easily be taken into account, there are no indications that they have extended their area in "jumps". The representatives of these two taxa are 0.0.1 and 0.0.2 species, with the one exception of the ascidian *Metrocarpa nigrum*. The two main migrant families of Decapoda are the Penaeidae and the Portunidae, both of which are active swimmers. Families of Decapoda which disseminate by means of long-lived planktonic larvae are not represented among the Lessepsian migrants. Barash and Danin (1972) mention that among Indo–Pacific Mollusca found in the Mediterranean the number of Lamellibranchia equals that of the Gastropoda; usually in the marine environments there is a preponderance of Gastropoda. It will be interesting to find out what the reason for that is, whether the mode of reproduction or bottom-type preference.

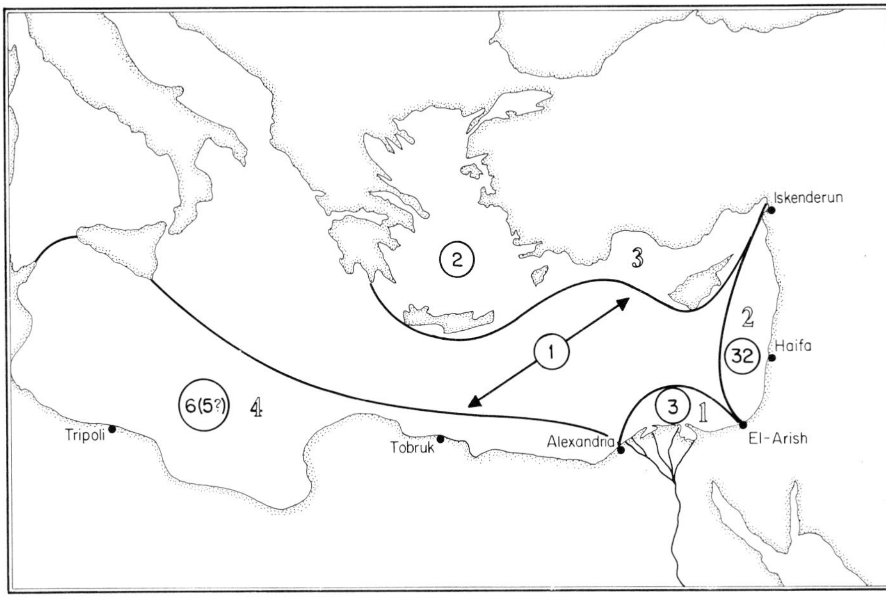

Fig. 40. Distribution of the migrant molluscs in the Eastern Mediterranean (legend as in Fig. 36) (original)

There are very few planktonic species among the Lessepsian migrants. In fact, we can count here only the two diatoms *Chaetoceros coarctatus* and *Rhizosolenia indica* and three species of Copepoda Calanoida: *Acartia centrura*, *Calanopia elliptica*, and *Calanopia media*. Lately, Dowidar (1973) reported the appearance in 1966 of *Ceratium egyptiacum* in the offshore area of El Arish. He considers the presence of this species in the Mediterranean as being dependent on continuous supply from the Red Sea. The planktonic decapod *Lucifer hanseni* has not succeeded in leaving the Canal opening yet.

The two or three problematic Hydrozoa among the Lessepsian migrants might have used passive transport rather than free planktonic medusae. For such outright holoplanktonic groups like Chaetognatha, Appendicularia, Thaliacea there is no known case of Lessepsian migrants, although the fauna of the Eastern Mediterranean is well known (Furnestin, 1953; Fenaux, 1971; Kimor, 1972; Godeaux, 1973, 1974).

There certainly are several animal groups in which information is scarce or even non-existent and therefore, they have no *known* Lessepsian migrants: for example, benthic Diatomacea, Actinaria, Platyhelminthes, Nemertes, Nematoda, Isopoda, Amphipoda, Ostracoda, Acarina. In some groups where we know of only 1–2 cases of migrants (e.g. Pantopoda, Hydrozoa, Cumacea, Tanaidacea, Polyplacophora), further investigation may add a few more Lessepsian migrants.

However, for the Scleractinia (the reef-building corals) and the Octocorallia, it is clear that no migration has occurred. The wealth of animal species—whole families and orders associated with the coral reefs—did not enter the Mediterranean either. In some well-investigated animal groups: Cirripedia (barnacles),

Table 7

Movement	Ecological Valency	Reproduction	Food
1. Active swimmers	4. Euryhalinity of the metahaline type (hyper-euryhalines)	7. Long reproduction season	10. Eury-phagy (non-selective feeding)
2. Possibilities of short-distance passive transport	5. Species with non-selective substrate requirements	8. Short-lived (lecithotrophic) planktonic larvae, or direct reproduction	
3. Free roaming benthic organisms	6. Eurythermy for lowered temperatures	9. Asexual reproduction	

Echinodermata, Cephalopoda, Selachia (sharks), migration did not occur or was minimal.

From all these, and also from some facts to be discussed in more detail below, we can draw the list of attributes that make a successful Lessepsian migrant. These fall under different categories (Table 7).

Since as a rule the migrational process includes a period of several generations of life within the Canal, properties necessary for the survival in the waterway are of highest importance. The ten biological properties listed above can eventually be subdivided into two categories, i.e. those which are obligatory for every migrant species and those which are alternative. It is evident that every Lessepsian has to fulfill at least three conditions: (a) to be euryhaline for high salinites; (b) to be nonselective as far as substrate is concerned; and (c) to be a euryphagous, nonselective feeder.

Eurythermy is probably useful but not necessary; it is probably the main reason for the distributional success in the expansion of the farthest advanced migrants (0.0.3 and 0.0.4 species). The different types of reproduction, or the three alternative ways of locomotion can appear in different combinations. One has to take into account that animals which are sessile in their adult phase are advancing by means of their short-lived pelagic larvae. Long-distance displacement on ship hulls—has led to cosmopolitan distribution; such species cannot be considered Lessepsian migrants.

Stenohalinity, monophagy and specialized substrate requirements are the main factors which exclude a species from Lessepsian migration. Holoplanktonic life or reproduction through long-lived planktotrophic larvae are the other important limiting factors. Scleractinians, for example, are stenohaline and need sediment-free rock substrates; most echinoderms are also stenohaline and have planktotrophic larvae; many fishes and gastropods are too monophagous to find a suitable new food organism in the Mediterranean.

It is an important task for future research to find out which are the specific inadequacies in the case of every species which has not used the Lessepsian opportunity. This critical adaptive unfitness can understandably also appear at some young reproductive stage. Such research is no doubt necessary in order to explain, for instance, why *Cassiopea andromeda*, the medusa which was so successful in colonizing the Canal, did not succeed in expanding into the Mediterranean.

The broad ecological valency of the Lessepsian migrants makes them very successful in the intraspecific competition in their new environment. Therefore, the biotic relationships established in the Mediterranean between the invaders and the autochthones probably—within certain environmental limits—lead to the success of the Lessepsian migrants.

Unlike some of the well-known invasional episodes of terrestrial zoology, the Lessepsian migration is a success story. The Lessepsian migrants are gradually building up their populations and expanding within their new marine province. I know of no case in which a Lessepsian migrant recorded at an early date disappeared afterwards. There are no reported cases of area restriction following the primary expansion. There were no population explosions and population crashes.

Unfortunately, the process of colonization, of niche conquest, or of species displacement has not been followed and studied even in one single Lessepsian species. The few scattered and circumstantial data available in order to describe the fate of the Lessepsian migrants within their new environments are compiled in Chapter 3.7.

A highly significant subject for investigation related to Lessepsian migration has only recently been approached: the study of the *parasite fauna* of the migrants can be of importance in three very important respects: (1) comparative parasitology may give additional clues, confirm or disprove zoogeographical conclusions; (2) it may explain the success of the migrant in its new environment; and (3) it may indicate newly established interactions between local and newcomer species and reveal a potential threat of epidemics.

Unfortunately, the only data available are those concerning ichthyoparasites, and these data are provided exclusively by the work done by Paperna and some of his collaborators. In Paperna's words (1972d): "Parasites may be introduced by migrant fish and consequently infect the autochthonous fish. The change in the environment may increase the pathogenic effect of the parasites on their original migrant host. The original parasitofauna of the migrant host may be lost altogether in the new environment."

Paperna (1972d) concludes that endoparasites cannot migrate unless their intermediary hosts, i.e. invertebrate species, actually migrate together with the fish species. Only the snail *Pirenella cailliaudi* which is known to harbor metacercariae of several species of helminths, might be a source for endoparasites alien to the Mediterranean.

Ectoparasites migrate together with their hosts if they are euryhaline and eurythermal enough. A "cleaning" of the fish through migration into a more saline environment has been documented by Paperna and Lahav (1975) for the *Mugil* species of the Sirbonic lagoon (Bardawil).

The ectoparasitic monogenean *Ancyrocephalus salinus* has followed its host *Aphanius dispar* everywhere: it is known from the Red Sea, the Mediterranean, and also from the landlocked springs around the Dead Sea (Paperna, 1964).

The clearest results have been obtained from Mediterranean populations of *Siganus rivulatus* and *S. luridus*. The two fishes have carried into the Mediterranean some of their Red Sea monogeneans: *Pseudohalitrematoides oramini* and *P. polymorphus* (Paperna, 1972d). On the other hand *Crenidens crenidens* the lagoon-dwelling fish has a different parasitofauna in the Mediterranean lagoons compared with those of the Red Sea. In the lagoons it shares the monogenean *Lamellodiscus elegans* with the Mediterranean fish *Diplodus sargus* and the same trematode species with the Mediterranean *Sparus aurata*. The euryhaline Mediterranean fish *Dicentrarchus punctatus* recently reported from the El Bilaiyim lagoon in the Gulf of Suez has a rather different parasitic fauna (Paperna, 1974). One may surmise, on grounds of comparative parasitology, that the two euryhaline fishes may have migrated in their opposite ways in pre-Lessepsian times, through estuarine, isthmian connections; they therefore had time to change their original parasitofauna. Some Lessepsian migrants, e.g. *Hemiramphus far* and *Sphyraena chrysotaenia* do not carry a Red Sea parasito fauna in their new Mediterranean habitat.

Paperna and Overstreet (in press) consider that some parasites of *Mugil* in the Mediterranean, such as the monogeneans *Benedenia* n. sp. and the copepod *Nipergasilus bora* may be regarded as Indo-Pacific immigrants. On the contrary, the acanthocephalan *Neoechinorhynchus agilis* and the copepod *Pseudocaligus apodus* may have migrated together with the mullets into the Red Sea.

It is clear that several species of parasites have to be added to the list of the Lessepsian migrants and possibly also to the "Anti-Lessepsians" (i.e. migrants *to* the Red Sea). These are evidently species which are at least as euryhaline as their hosts.

There is no sense to complain again on the lack of further information in this interesting field of study. Suffice it to quote from Paperna (1972d) a questioning sentence which still needs an answer: "If immigrant species are likely to be relatively less infected with parasites and pathogens, will they eventually become more competitive and eventually overtake the local species?"

3.7 Analysis of the Colonizing Success of Lessepsian Migrants

Very little has been done to understand the reason for the success of the Lessepsian migrants. We have no adequate monitoring of the buildup of the migrant populations, of their dependence on the environmental factors, and of their interactions with the autochthonous biota.

In most cases we are confined to occasional collecting data, from scattered localities. In very few cases have the findings been rechecked in the same locality. This accidental character of our knowledge—evident at the qualitative level discussed above—becomes more painfully felt if we try to reach quantitative, dynamic or ecological conclusions.

One circumstance is nevertheless clear and beyond doubt: the Lessepsian migrants are successful species, expansive *r* strategists (Pianka, 1970) and there are no known examples of a reversal—i.e. of a disappearance or radical depletion of the Lessepsian population in the Mediterranean. In the words of Gilat (1969): "We have evidence that all the immigrants became members of the existing communities, some of them even occupied dominant positions ... There is evidence that there is no regress in the building up of populations of the new immigrants".

Some data exist from three ecological categories of migrants, namely: demersal fish; infralittoral level-bottom macrofauna; and intertidal rock-dwelling fauna.

The role played by the fish among the Lessepsian migrants has been shown to be very important. According to Ben-Tuvia (1972), 11.6% of the fish species along the Israel coast are immigrants from the Red Sea. No less than 21 of the 27 species of High Probability Lessepsian migrants appear regularly in commercial catches. According to George and Athanassiou (1966), 5 of the 26 commercially important fishes of the beach seine catches of the Gulf of St. George in Lebanon are immigrants.

The analysis of the catch statistics of the Israel fisheries gives very interesting information concerning a few migrant species such as *Upeneus moluccensis*, *Upeneus asymmetricus*, *Saurida undosquamis*, *Leiognathus klunzingeri*, *Sphyraena chrysotaenia*, *Atule djeddaba*, *Dussumieria acuta*, *Pranesus pinguis*, and *Hemiramphus far*. While the pelagic fishery landings are mostly composed of the autochthonous species *Sardinella aurita* (Ben-Tuvia, 1973), the pelagic Lessepsian migrants play chiefly a role of food basis for the big open sea predators. Ben-Tuvia (1973) quotes the work done by Bograd-Zismann between 1963–1970 on the food chain of the little tuna *Euthynnus alleteratus*. Several Lessepsian species occurred in the stomach contents of this species, especially in younger age groups. *Hemiramphus far* made up 3% of the food of older specimens of *Euthynnus*. Another species, *Pranesus pinguis*. was an important food item in the month of January.

Among the trawl catches of demersal fish there are several species of migrants of the highest economical value (here the statistics enabled some analytical conclusions). Especially important are the indications concerning direct or indirect competition between species.

The first pair of species which has been investigated are from the Mullidae (goat fishes): the local *Mullus barbatus* (the red mullet), and the immigrant *Upeneus moluccensis* (the yellow-striped goat fish). Wirszubski (1953) considered that 10–15% of the catch of Mullidae in the years 1946–47 and 1949–1950 was made up of the immigrant species. According to Oren (1957), following an exceptionally warm winter the catch of *Upeneus moluccensis* was 83% of the mullet catches of 1956. Since 1954 the Lessepsian migrant species has maintained a more or less permanent percentage of 30% (Ben-Ţuvia, 1973). There are indications that temperature plays an important role in the quantitative relationship between *Mullus* and *Upeneus*. It is interesting also to note (Ben-Tuvia, 1973) that *Mullus* remained consistently dominant on bottoms deeper than 70 m.

A similar relationship was studied by Ben-Yami and Glaser (1974) and is the first attempt to reach conclusions about the niche competition between a local species and an invader: The invader *Saurida undosquamis* was first reported in the Mediterranean in 1952; in 1953 it appeared in trawl catches in the El Arish–Gaza area (Ben-Tuvia, personal communication).

In the winter of 1954–55 the amount of lizard fishes increased dramatically and since then it has constituted 20% of the trawl catches and 11% of the total landings of the Israel sea fisheries (Ben-Yami and Glaser, 1974). *Saurida* does not compete with the local Atlantic lizard fish *Synodus saurus* which is rare and noncommercial, but with the hake *Merluccius*. The two fishes share the same food and frequently the same bottoms. The above-mentioned authors reach the conclusion that *Saurida* takes advantage of dry warm winters and *Merluccius* of the rainy cold winters. Ben-Yami and Glaser (1974) also reach the conclusion that the two species have divided the shelf between them, in such a way that in the summer the lizard fish predominates shallow bottoms and is barely overlapping with the hake, while in the winter, the hake advances into the shallow areas. *Saurida* is generally caught down to maximum 80 m. *Merluccius* appears in the summer chiefly below this depth. It seems therefore that *Saurida* might have occupied the shallower parts of a depth range previously occupied only by the hake. The food of the two species is rather similar, composed mainly of anchovy, and lately of two Lessepsian migrant fish: *Leiognathus klunzingeri* and *Upeneus asymmetricus*. Ben-Yami and Glaser (1974) reach the conclusion that the sudden appearance of *Saurida* did not increase the trawl catches but merely displaced *Merluccius* to deeper bottoms. According to Gorgy (1966), the hake is found along the Egyptian shores only deeper than 100 m. Since the 1954–56 invasion, *Saurida* has maintained its numbers and has not descended below 13% of the total trawl catches (Fig. 41).

Ben-Tuvia (1973) mentions that before the appearance of *Saurida* in commercial amounts in 1955, the catches of the immigrant food fish *Leiognathus klunzingeri* were bigger. This conclusion would in a way contradict the previous conclusion about the penetration of *Saurida* into the same food niche. An additional prey such as *Leiognathus* may have reasonably increased the available food basis for the two competing species *Saurida* and *Merluccius*.

Speaking of the near-shore fishes, Ben-Tuvia (1973) mentions that *Siganus rivulatus* may make up as much as 50% of the fishes caught in fish poisonings. Popper (personal communication) considers that *S. rivulatus* thrives at the expense of the local fish *Sargus*. Among the barracudas the immigrant *Sphyraena chrysotaenia* makes up presently one half of the catches, while the two local species *S. sphyraena* and *S. viridensis* constitute the other half (Ben-Tuvia, 1973). Since here too the local species have not disappeared, the logical explanation is that the migrants fitted themselves into the food niche of the natives, by better exploitation of the available food or due to the enrichment in the food basis itself by the addition of migrant species.

In the case of the immigrant shrimp and portunid swimming crabs immigration has resulted in a better exploitation of a preexistent food basis. There are six species of commercial shrimp in the Levant basin; four of them are Lessepsian migrants: *Penaeus japonicus*, *P. semisulcatus*, *Metapenaeus monoceros* and *M. stebbingi*; two of them *Penaeus kerathurus* and *Parapenaeus longirostris* are Atlantic species. According to Tendler (personal communication) the migrant shrimp make up 80% of the Israel shrimp-catches. Along the Anatolian coast another Lessepsian migrant shrimp becomes commercially important: *Trachypenaeus curvirostris* (Geldiay and Kocatas, 1972). According to these authors *Penaeus japonicus* is usually more abundant at Iskenderun than *P. kerathurus*.

If one analyzes the tables of the bathymetric distribution of benthic invertebrates as given by Gilat (1964, 1969) for four profiles along the coast of Israel and

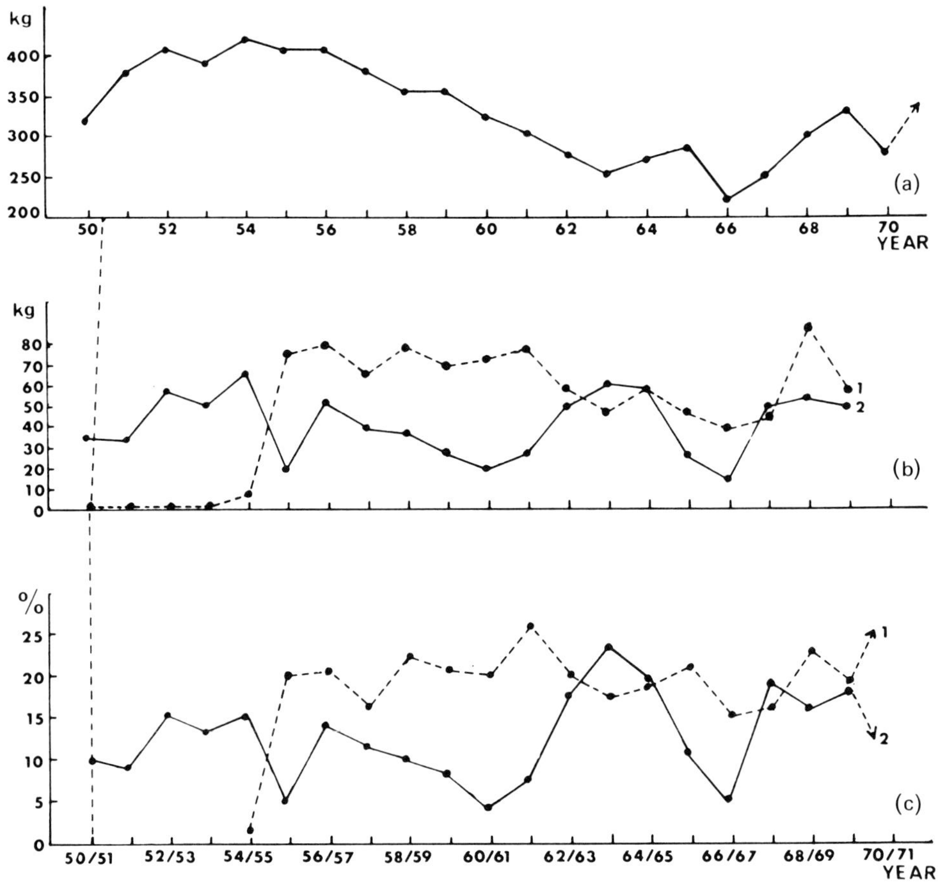

Fig. 41 a–c. Trawl catches and ratio of *Saurida* and *Merluccius* in Israel Mediterranean area (1950–1970). (a) all fishes in catch per unit effort; (b) same for *Saurida* and *Merluccius*; (c) proportion in catch of the two species. *Saurida*: full line; *Merluccius*: dashed line (from Ben-Yami and Glaser, 1974)

North Sinai (Rubin, 1960–1962; Alexander, 1968–1969; Ashdod, 1968–1969; El Arish, 1968–1969) it appears that the autochthonous *Parapeneus longirostris* extends down to over 200 m, while the Indo–Pacific species are confined to bottoms shallower than 80 m. The numbers of *P. longirostris* also increase towards greater depths.

Speaking of benthic crustaceans, it is interesting to observe the congeneric pair of mantis shrimps: the Atlantic *Squilla mantis* and the immigrant *Squilla massawensis*. Although known from Alexandria in the late thirties, *S. massawensis* was reported from the coasts of Israel only in 1955. Since then, it has appeared constantly together with *S. mantis*. According to the data from Gilat's profiles (see below) *S. massawensis* is chiefly collected at depths of 10–50 m, *S. mantis* between 35 and 80 m.

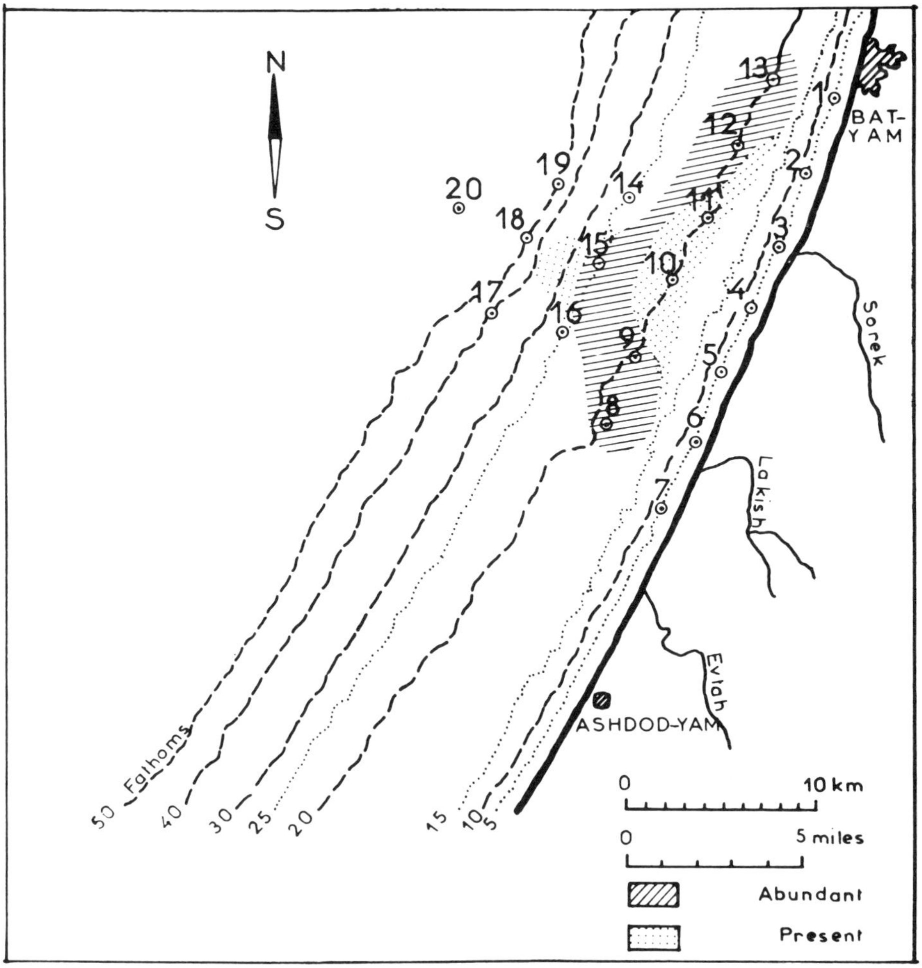

Fig. 42. Bathymetric distribution of *Charybdis helleri* (Decapoda) off southern coast of Israel (from Gilat, 1964)

The success of the colonizing crabs was remarkable. The swimming crab *Charybdis longicollis*, reported for the first time in the Gulf of Mersin in 1954, became in the later years a "true pest" in the words of Lewinsohn and Holthuis (1964). Gilat (1964) found this species as a dominant benthic form on the sedimentary bottoms off Rubin between 1960–1962. *Charybdis longicollis* constitutes up to 70% of the macrobenthic biomass on bottoms of mixed sand and mud. This species was found to be most abundant around 30 m depth and disappeared at depths greater than 80 m (Fig. 42). *Charybdis* also maintained its plentiful populations later (Gilat's 1968–1969 profiles) and also along the whole geographical range from El Arish profile to Alexander profile, and was not found below 65 m depth.

Two species of swimming crabs, the Lessepsian migrant *Portunus pelagicus* and the recent West Atlantic intruder *Callinectes sapidus*, are fairly common in the catches from depths of 15–50 m along the Egyptian and Sinai Coast (Gulf of Tina) (Gorgy, 1966). For all these species of portunid crabs it is difficult to indicate a competing local species which has been displaced. None of the local portunids was frequent enough to be compared with the wealth of the three species of newcomers.

The success of the leucosid crab *Myra fugax* reported from Jaffa as early as 1930 cannot be regarded as a competitive displacing of a closely allied local species. In the Rubin profile (Gilat, 1964) it appears as a dominant species besides *Charybdis longicollis*, on the same type of bottoms between 35–50 m (Fig. 42). It seems however that *Myra fugax* is still frequent at depths over 100 m, probably the deepest going Lessepsian migrant, encountered also on bottoms of clean mud. There are two species of local leucosid crabs in the Eastern Mediterranean., *Ebalia granulosa* and *Ilia nucleus*, but they are not known to be frequent in any of the investigated stations. Another Indo-Pacific leucosid crab, *Leucosia signata*, has also appeared recently in the Levant.

There are few data about the colonizing success of the Lessepsian migrants among the level bottom molluscs. According to Gilat (1969) the bivalve *Paphia undulata* is a fairly characteristic species of mixed bottoms between 10–35 m. It does not appear on shallower clean-sand bottoms and not deeper than some 35 m. A curious case of sudden massive appearance is that of the gastropod *Cerithium kochi*. This Red Sea species, not reported in the Suez Canal, did not appear in Gilat's collections from Haifa Bay (1953–1956) or from Rubin (1960–1962). First dredged in 1963 in the Bay, it has since appeared rapidly as one of the dominant species of the level bottoms of the Israel coast–between 20–60 m depth. In Gilat's 1968–1969 profiles it was already one of the mass organisms. It is very difficult to assume that this sudden explosion of *Cerithium* populations, maintained at about the same numerical levels, has occurred at the expense of some other species. There is to our knowledge no species of gastropods which has such huge biomasses on the inshore level bottoms of Israel.[1]

It is interesting to analyze the qualitative importance of the Lessepsian migrants among the general species diversity which characterizes the different depths of the El Arish profile (Gilat, 1969).

Between	5–8 fathoms	1 Lessepsian out of 9 frequent species
Between	10 fathoms	5 Lessepsians out of 10 frequent species
Between	12 fathoms	7 Lessepsians out of 12 frequent species
Between	20–25 fathoms	3 Lessepsians out of 18 frequent species
Between	30–40 fathoms	1 Lessepsian out of 19 frequent species
Between	50–100 fathoms	2 Lessepsians out of 30 frequent species

[1] *Note added in proof:* Recently, M. Tom in a thesis "The Benthic Fauna Association of Haifa Bay" (Tel-Aviv University, October 1976), compared the benthic diversity of samples collected in 1974–75 with those collected there by Gottlieb-Gilat in 1953-56. Four Indo-Pacific species have been found as new and numerous members of the benthic communities of Haifa Bay: the crab *Charybdis longicollis* was frequent on biogenic rocks and the gastropods *Cerithium kochi* and *Isanda* cf. *holdsworthiana* on soft bottoms between 25–30 m. The brittle star *Ophiactis parva* (Mortensen), a new addition to the list of the Lessepsian migrants, has been found by Tom as a frequent inhabitant of the *Caulerpa scalpelliformis* meadows and on the arms of *Antedon mediterranea* around 40–50 m depth.

Turning now to the littoral rock dwellers, information becomes increasingly scarce. Safriel and Lipkin (1975) compared the repartition of 222 species along the Intertidal zonation pattern of the Mediterranean coast at Mikhmoret—201 native species and 21 Lessepsian species. They found no immigrants in the Supralittoral, fewer in the Upper Midlittoral than proportionally expected, normal proportions in the Mid Midlittoral and Lower Midlittoral and a relatively higher proportion than expected in the Infralittoral. Several of Safriel's students are at present studying the interspecific relationship between native and colonizing species in the rocky littoral: Felsenburg (1974) could not find any clearcut difference between the present vertical distribution of the autochthonous *Mytilus minimus* and the migrant *Brachidontes variabilis*. These two species appear interspersed, and *B. variabilis* did not change its vertical distribution in the intertidal belt after entering the Mediterranean. Ayal (unpublished) is studying the competition assumed to exist between the colonizing species *Cerithium scabridum* and the non-colonizing autochthonous *Cerithium rupestre*. The two species occur together along the shores of Israel, north of Caesarea. Ayal has found *C. scabridum* to be dominant at present, sometimes with thousands of specimens per square meter, in a site in the Bay of 'Akko (Acre) where in 1964, according to Safriel, *C. rupestre* was dominant. According to Ayal there is an explosive settling of young specimens of *C. scabridum* in September–October but during the winter the cold frequently destroys the whole population of settlers. It is important to mention that *Cerithium scabridum* lives in the shallowest tens of centimeters of water where the cooling in the winter is most drastically felt.

The Bay of Acre is an especially favorable site for the Red Sea immigrants. The bay is sheltered from the open sea by the additional structures of the old Crusader and Turkish port; pollution is minimal since the harbor is used only by small fishing craft. The waters are quiet, undisturbed and warm. This environment shows at present the most abundant populations of the pearl oyster *Pinctada radiata* on the Israel coasts: the pearl oyster forms whole reef-like structures. It is here that huge concentrations of the red-colored immigrant crab *Atergatis roseus* are found, a species first reported in the Mediterranean in 1961.

The little sea star *Asterina wega* is very frequent at 'Akko (Acre). Achituv (1969, 1973) has studied the appearance of this Lessepsian migrant in the Mediterranean and the parallel disappearance of the autochthonous *Asterina gibbosa*. *A. wega* was first collected in small numbers in Haifa Bay, in 1955. Since then it underwent a regular population explosion and is found by the thousands. Achituv (1973) collected 530 specimens of this little sea star at Akhziv on one single occasion in June 1970.

It seems that in some relation with the appearance of *A. wega*, the other species *A. gibbosa* disappeared. Out of the 37 specimens of *A. gibbosa* found by Achituv in the collections of the Hebrew University, 34 were sampled at 'Atlit between 1934 and 1944. In 1970 Achituv was unable to find any *A. gibbosa* on the collecting sites of 'Atlit whence most of the known specimens came. The locality, however, yielded specimens of *A. wega*. It seems that *A. gibbosa* was last collected at Shiqmona in 1963. On the island of Cyprus, according to Achituv, *A. gibbosa* alone is found, and frequently.

Two facts need to be emphasized: *A. wega* is a species which reproduces asexually by fission and subsequent regeneration, and Achituv found little indication that the species reproduces sexually in the Mediterranean. This is a circumstance which was no doubt most favorable for the spreading and colonizing success of *A. wega*. Besides, this species is now found only north of 'Atlit, and the various populations show clear differences in sizes and in the regeneration coefficient. One may assume that a favorable year, or period of years, has enabled several small populations ("propagula") to proliferate rapidly and independently from a general gene pool.

The nearshore waters offer several other examples in which the biology of colonizing can be investigated: for instance, the interrelation between the local purple snail *Thais haemostoma* and the Lessepsian migrant *Thais carinifera* first reported in 1956 from Tel-Aviv. The migrant species has been fairly common since and the analysis of the relation between the two species is easy to follow because they leave behind clearly distinguishable egg capsules. In two other molluscs common along the rocky shores of Israel—the key-hole limpet *Diodora rüppelli* and the hammer oyster *Malleus regula*—there is no obvious evidence that their settling success was detrimental to any local species.

Finally, the meiobenthos offers another interesting case of niche occupation by a Lessepsian migrant species. I have shown (Por, 1964) that the bathymetric succession of the Copepoda Harpacticoida of the Canuellidae is different, compared with European waters, because of the appearance of the Lessepsian migrant *Scottolana longipes*. Usually *Canuella perplexa* occupies the predominantly sandy bottoms and *Canuella furcigera* follows deeper on the predominantly sandy muddy bottoms. Along the Levant shores, *Scottolana longipes* clearly occupies the belt in which usually *C. furcigera* is found. It is interesting that *S. longipes* descends to such depths as 430 m and occurs everywhere from 20 m downwards in great numbers.

A few conclusions can be drawn from the analysis of the relatively scarce data that we have:

1. The colonization by the Lessepsian migrants is not of the type studied in various cases of invasions of terrestrial animals. There is a build-up of populations, sometimes even explosive, but this is not followed by a population crash but by constant and relatively high numbers.

2. Following Gilat's words (1969), it looks as if the Lessepsian migrants first have small populations, propagula, which are "waiting" for the favorable environmental episode to build up a significant population.

3. In agreement with Ben-Tuvia (1973), I consider temperature to be the most important single factor in triggering the build-up of a successful migrant population, for example the many successful populations of Lessepsian migrants which appeared after the warm winters of 1955–1956 and 1960–1961.

4. The Lessepsian migrants are most successful at intermediary depths, between 20–40 m on mixed sandy–muddy bottoms. The rock dwellers prefer the infralittoral. The reason is the strong cooling of the shallow waters in the winter and the exposure to cold at higher intertidal levels. Usually, the depth limit of the Lessepsian migrants is 70–80 m: this is the depth which according to Oren (1970), has a year-round temperature limit of 18° C.

Table 8. Congeneric or other interspecific species-pairs in the Eastern Mediterranean

Local Mediterranean species		Red Sea species (Lessepsian Migrant)
Canuella furcigera	3.3.2	*Scottolana longipes*
Parapenaeus longirostris *Penaeus kerathurus*	2.2.3 3.2.3 2.2.3 2.2.3	*Penaeus japonicus* *Penaeus semisulcatus* *Metapenaeus monoceros* *Trachypenaeus curvirostris*
Ilia nucleus *Ebalia granulosa*	2.2.3	*Myra fugax*
Squilla mantis	2.2.2	*Squilla massawensis*
Mytilus minimus	1.1.4	*Brachidontes variabilis*
Cerithium rupestre	1.1.2	*Cerithium scabridum*
Thais carinifera	2.2.2	*Thais haemostoma*
Asterina gibbosa	2.3.2	*Asterina wega*
Sphyraena sphyraena *Sphyraena viridensis*	3.3.4	*Sphyraena chrysotaenia*
Mullus barbatus	3.3.2 3.2.4	*Upeneus moluccensis* *Upeneus asymmetricus*
Merluccius merluccius *Synodus saurus*	4.2.3	*Saurida undosquamis*

5. A second reason for the typical depth preference of the Lessepsian migrants can be seen in their nonselectivity for substrate (see previous chapt.). This gives them an advantage on mixed sand–muddy bottoms and on rock bottoms prone to be covered with sediment.

6. Other than the case of *Asterina gibbosa* there is no known case in which a Lessepsian migrant species has completely replaced a local one. The rule seems to be a bathymetric readjustment between the local species and the invader, the local species maintaining its dominance on deeper, i.e. cooler bottoms. Examples are: *Merluccius*, *Squilla mantis*, *Parapeneus longirostris*.

7. Only in the case of *Saurida* vs. *Merluccius* do we have proof of competition for the same food organism. In all the other cases (with the exception of *Asterina wega* versus *A. gibbosa*) the immigrant accommodates itself very well with its supposed local competitor (Table 8). It seems as if the carrying capacity of the environments of the Levant allows for additional links in the food chains. In some cases such as the Penaeidae and Portunidae, or *Cerithium kochi*, we have instances of population build-ups on previously less populated bottoms. The case of *Brachidontes variabilis* and *Pinctada radiata* may be similar. The abundance of the prey fish *Leiognathus klunzingeri* may indicate that Lessepsian migrants increase the trophic basis of the Levant Basin.

The fact that at least 41 species of a total 204 Lessepsian migrants have become very common in the Levant basin is clear proof of the success of the immigration. From the 41 species only 6 or 7 are unknown in the Suez Canal. There seems to be no obvious correlation between early entrance into the Canal or early emergence into the Mediterranean, and the success of the Lessepsian migrants to establish big populations, along the Levant coast.

Five sketch maps are presented to illustrate the advance of five species of successful migrants: *Portunus pelagicus* (Fig. 43), *Cerithium scabridum* (Fig. 44), *Pranesus pinguis* (Fig. 45), *Siganus rivulatus* (Fig. 46), *Holocentrus ruber* (Fig. 47).

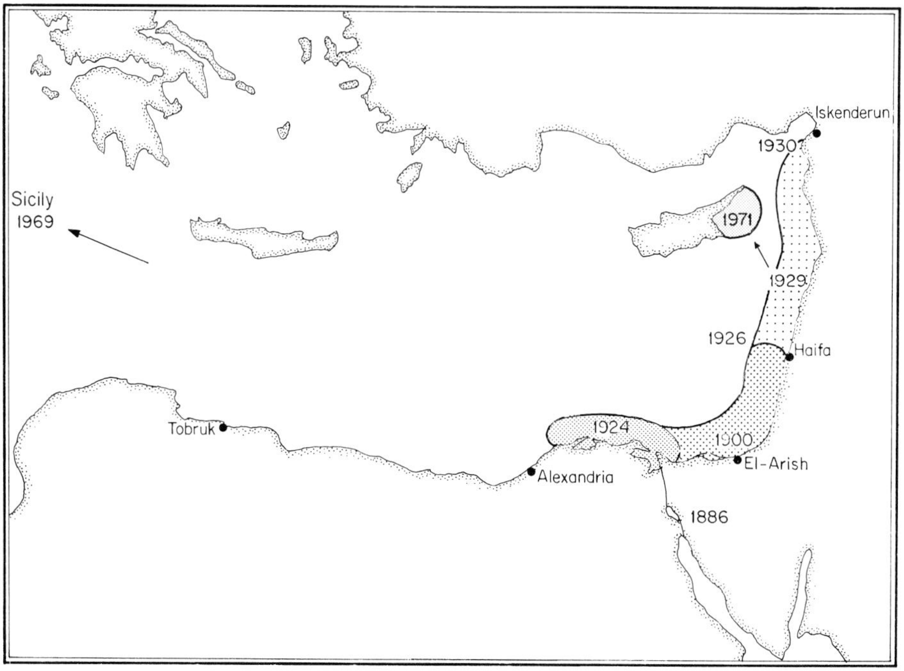

Fig. 43. Gradual advance of *Portunus pelagicus* (Decapoda) in the Mediterranean (original)

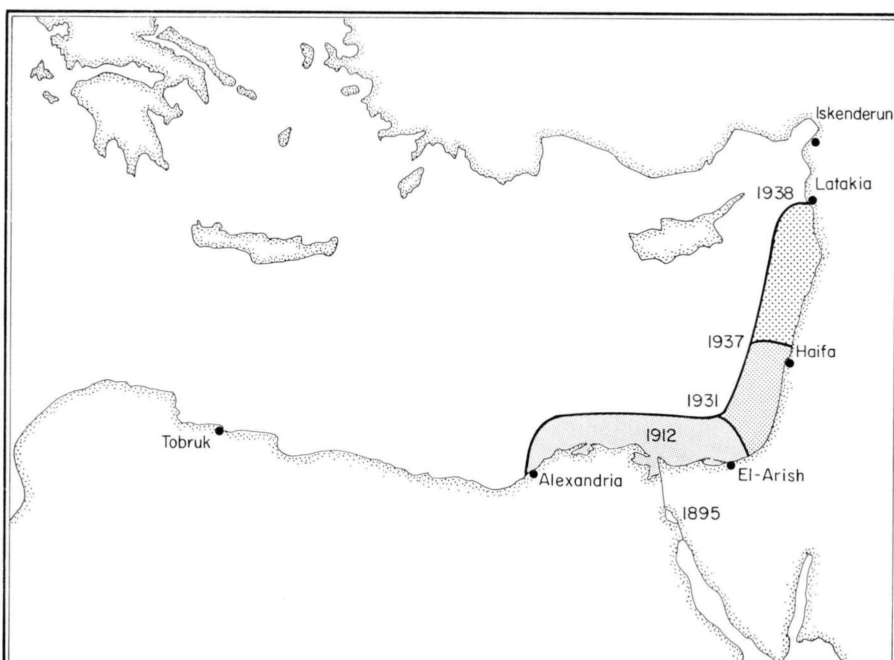

Fig. 44. Gradual advance of *Cerithium scabridum* (Gastropoda) in the Mediterranean (original)

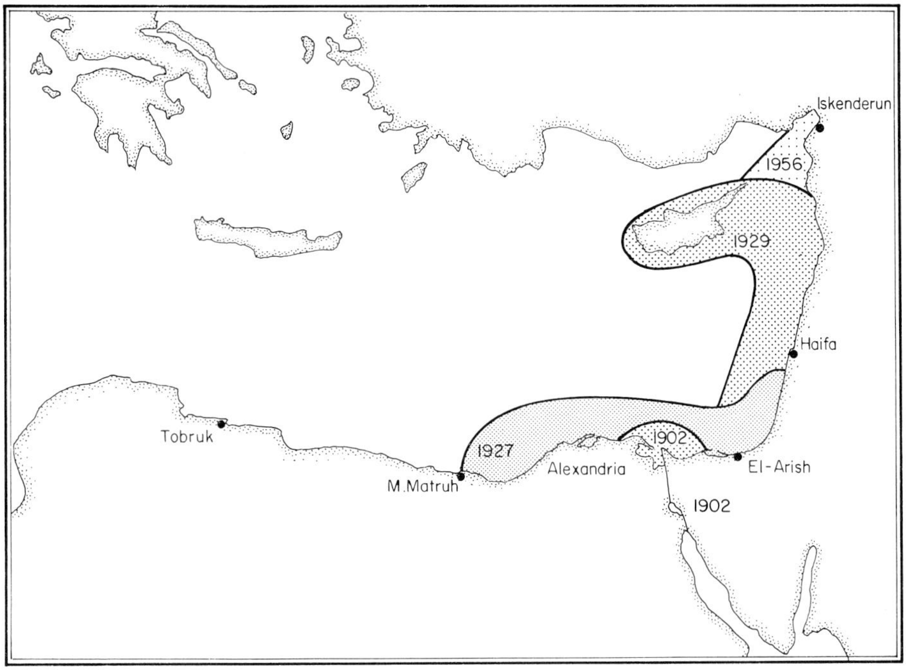

Fig. 45. Gradual advance of *Pranesus pinguis* (Pisces) in the Mediterranean (original)

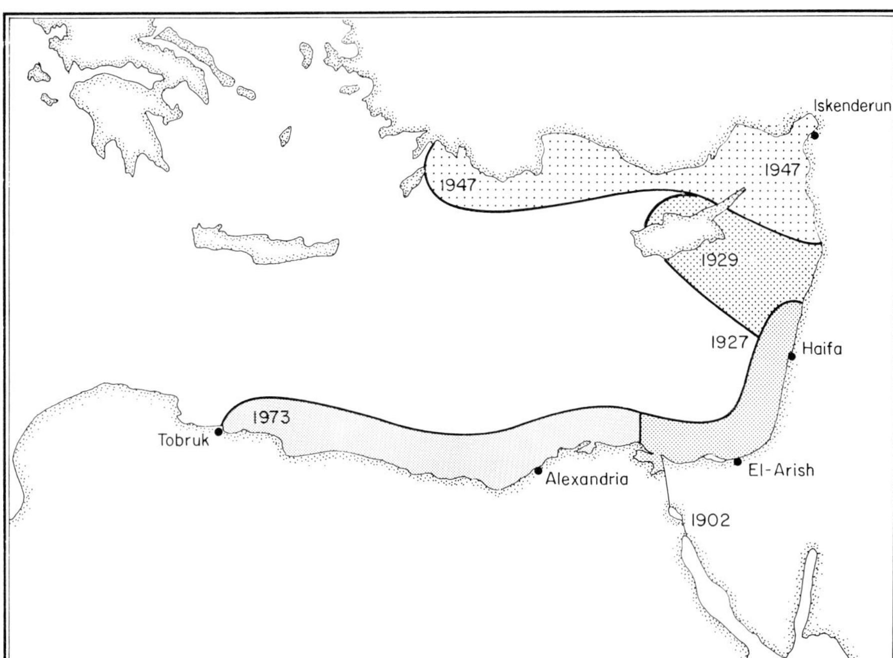

Fig. 46. Gradual advance of *Siganus rivulatus* (Pisces) in the Mediterranean (original)

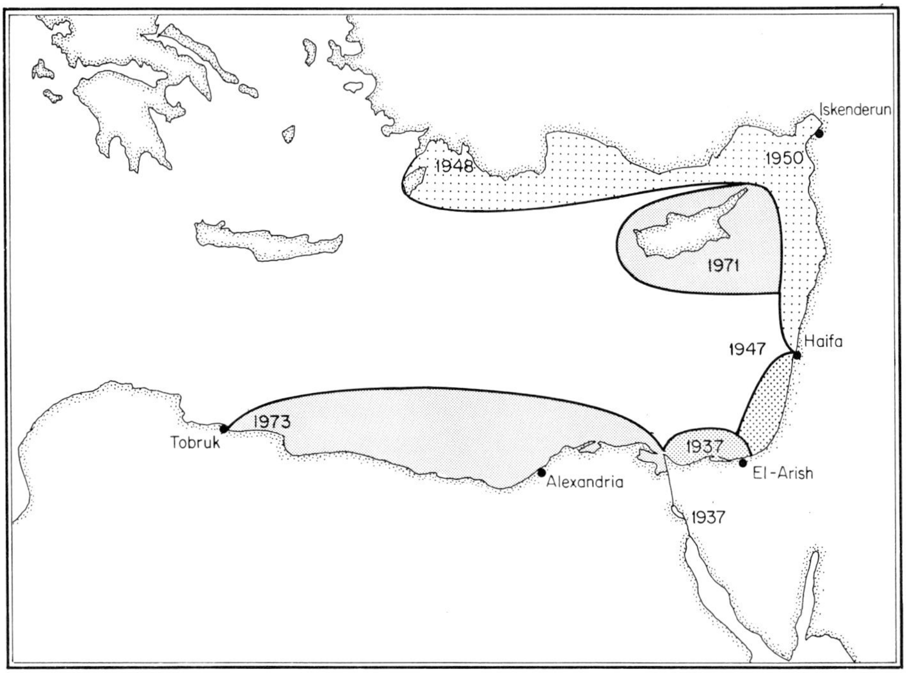

Fig. 47. Gradual advance of *Holocentrus ruber* (Pisces) in the Mediterranean (original)

In addition to the species tabulated above, the following Lessepsian migrant species are known to have considerable populations in the new environment:

Table 9. Very common Lessepsian migrants in the Levant Basin with unknown local competitor

4.1.2	*Caulerpa scalpelliformis* (alga)
2.1.3	*Halophila stipulacea* (see-grass)
4.2.2	*Chrotella cavernosa* (sponge)
2.1.2	*Pseudonereis anomala* (polychaete)
4.3.2	*Apseudes intermedius* (tanaid crustacean)
4.3.2	*Atergatis roseus* (decapod crustacean)
4.2.3	*Charybdis longicollis* (decapod crustacean)
1.1.3	*Portunus pelagicus* (decapod crustacean)
3.3.2	*Gastrochaena cymbium* (lamellibranch mollusc)
1.2.3	*Malleus regula* (lamellibranch mollusc)
1.1.4	*Brachidontes variabilis* (lamellibranch mollusc)
1.1.4	*Pinctada radiata* (lamellibranch mollusc)
3.3.2	*Bursatella leachi savigniana* (gastropod mollusc)
4.3.2	*Cerithium kochi* (gastropod mollusc)
1.2.2	*Diodora rüppelli* (gastropod mollusc)
4.3.2	*Hippopodina feegeensis* (bryozoan)
2.1.2	*Herdmania momus* (ascidian)
2.2.2	*Callionymus filamentosus* (fish)
1.2.3	*Dussumiera acuta* (fish)
1.1.4	*Pranesus pinguis* (fish)
1.1.2	*Atule djeddaba* (fish)
1.2.4	*Leiognathus klunzingeri* (fish)
1.3.4	*Siganus luridus* (fish)
1.1.4	*Siganus rivulatus* (fish)

3.8 Zoogeographical Consequences and the Future of Lessepsian Migration

First of all I shall try to summarize the zoogeographical impact of the Lessepsian migration as it now stands. It is important to be able to give a mathematical expression of the new situation created in the Mediterranean fauna after a century of Lessepsian migration. In the discussions above, more than 200 species of Lessepsian migrants have been mentioned. Taking into account the incompleteness of our knowledge, I would suggest 500 as the probable number of Lessepsian species actually present in the Mediterranean. Riedl (1963) estimates a total of 15,000 species of animals in the Mediterranean. If so, then Lessepsian migration has added about 3% to the Mediterranean faunal inventory.

Fredj (1972, 1974) analyzes the distribution of 1244 species of Mediterranean benthic invertebrates. He chooses a number of phyla rather arbitrarily; for instance, he refers to Lamellibranchiata, but not to Gastropoda. Fredj mentions only 40 Lessepsian migrants from among the Echinodermata, Polychaeta, Decapoda, and Lamellibranchiata. These constitute 3.21% of his list. Fredj himself speaks of 4.5% of the species which have a Mediterranean–Indo–Pacific distribution; however, his criteria are not similar to those used by me.

Including the other 37 Lessepsian migrants not mentioned by Fredj in the four taxa he deals with, it totals 77 out of 1281 species (1244+37). That brings the weight of the Lessepsian migrants to 6.01% of the taxa treated by Fredj.

Let us now use the figure of 48.8% given by Fredj for the percentage of those species out of the 1244 which live in the Eastern Mediterranean. The Eastern Mediterranean is by about a half poorer in species than the Mediterranean taken as a whole. Returning to Riedl's figure, one can, therefore, assume that some 7500 species live in the eastern part of the Sea, out of which the approximated number of 500 Lessepsian migrants make up 6.6% of the total.

However, the term Eastern Mediterranean, as generally used, is too broad, and sometimes includes the Aegean Sea. Considering the Levant Sea alone, the proportion of the Lessepsian migrants is even higher. Ben-Tuvia (1973) calculates that the immigrants constitute 11.6% of the fish-fauna of the Israel coast. Out of the Decapoda known in the Levant Basin the Lessepsian migrants form about 21% (Holthuis and Gottlieb, 1958; Lewinsohn and Holthuis, 1964). The percentage is perhaps of the same order of magnitude in Polychaeta and probably lower in Mollusca.

It seems to me justified to suggest that at present the Lessepsian migrants constitute about 10% of the biota of the Levant Sea. Does this percentage justify the separation of a special zoogeographical subdivision, a "Lessepsian province" or subregion within the Atlanto-Mediterranean region? The Black Sea, for instance, which has a Ponto-Caspian admixture of species to a majority of Mediterraneans, has frequently been considered as a separate zoogeographical province. There are, however, no generally accepted criteria for deciding at what point of the similarity–dissimilarity relation two areas should be considered different, and at what level. Banarescu (1970) suggests that "regions" are characterized by different families and subregions or provinces by vicariant genera. Ekman (1940) proposed his formula for comparing two faunas:

$$d = \frac{\text{Specificity of area A} + \text{Specificity of area B}}{\text{Similarity of A and B}}.$$

The different hierarchic taxa have, according to Ekman, different "species–age values": e.g. one species–age value for a difference on the species level, three species–age values for generic-level difference, 18 species–age values for family-level differences.

This formula works on the assumption that we are dealing with two faunas which have evolved separately, and the amount of diversity between them reflects a difference in their history. However, in the case of Lessepsian migration we are witnessing a sudden change, i.e. there is no vicariance involved but rather an addition of species to the preexisting inventory.

Nevertheless, this excursion into the field of Ekmanian quantitative zoogeography can be of use in one point. The species of Lessepsian migrants represent perhaps 10% of the Levant biota; however, they contain many genera and even numerous families which were alien to the Atlanto-Mediterranean region. To take the example of fishes, the Holocentridae, Leiognathidae, Theraponidae, Platycephalidae, Tetraodontidae and Monacanthidae were not previously represented in the Mediterranean fauna. Among the Mollusca there are also several cases of new families. Among the Decapoda there are no extra–mediterranean families, but many genera (*Metapenaeus*, *Myra*, *Leucosia*, *Charybdis*, *Atergatis* etc.). It seems beyond doubt that the area occupied in the Mediterranean by the Red Sea migrants can be considered as a Lessepsian province or subregion within the Mediterranean.

What the geographical limits of this province will be and the final proportion of the Lessepsian taxa in it is a matter for the future, on which only a few conjectures can be made. Although the great majority of the Lessepsian migrants are at present concentrated along the Levant coast, it is reasonable to assume that in due course they will occupy the whole of the Levant Basin and the Sidra Sea (the sea situated between the Tunisian and Cyrenian coast). The westward migration along the African coast was probably delayed by the opposing current, the structure of the outlet at Port Said which is isolated westward by a 6.5-km long pier, and perhaps also by the Nile water influx. Deficient information may also in part be the reason for the apparent paucity of Lessepsian migrants on the African shores (Stirn, 1973). There is, however, little doubt that the expansion of the migrants in the Mediterranean is a time–distance function within frames set by temperature limits. This temperature limit has to be sought at the 16° C isotherm of minimum winter surface temperature. Therefore, most of the Eastern Mediterranean, excluding the Aegean Sea proper and the Adriatic Sea, has to be considered within the radius of Lessepsian migration.

Indeed, there are differences between the power of expansion of the various migrant taxa. Fishes can appear seasonally far westward during the summer and retreat to warmer waters in the winter. To some extent, different decapods can develop similar migrational patterns. Molluscs, however, are tied to their substrate, and therefore show also less tendency to explore the periphery of the Lessepsian distribution radius. However, some molluscs, e.g. *Brachidontes*, *Malleus*, *Pinctada* may reach protected and warm gulfs or portuary areas in the West by passive transport.

One can also assume that the depth distribution, limited along the Israel coast by the 70–80 m isobath, is gradually changing towards the west and the Lessepsian migrant species are restricted to a more and more limited bathymetric range; a decreasingly thick sublittoral belt is still able to provide the more or less uniform and warm temperatures required.

Since the colonizing success of the immigrants within the local communities is also influenced by the temperature sensitivity of the intruder, versus that of the local competitor, the chances of success will decrease from East to West, with the lower isotherm values. The very carrying capacity of the environment is also decreasing westward, being presumably inversely proportional to the increasing species diversity of the local Mediterranean fauna.

This seems to be a fairly static picture. Several authors, chiefly Tortonese (1951) and Kosswig (1956), have assumed a rapid physiological evolution of the Lessepsian migrants which leads to speciation, or hybridization with local congeneric species. I myself have raised this possibility in an introductory chapter of this book, and the eventuality of such phenomena cannot be excluded. However, nothing in the direction of an evolutionary divergence of the Lessepsian migrant species could until now be externally and descriptively observed. This is also supported by Tortonese (1973 b) regarding fishes and echinoderms. There are differences in size between the specimens of the Red Sea population of origin and those of the "Lessepsian diaspora". According to Ben-Tuvia (1966) there is a shift in the reproduction period of some migrant species of fish. These differences however, belong very much to the infra-subspecific level.

It is regrettable that no experimental work has as yet been done to check the degree of interfertility of Lessepsian migrant populations with their populations of origin. It is also possible that differences in tolerance of preference versus environmental factors can already be experimentally discerned when two conspecific populations are compared. Evolution of new taxa even on the subspecific level, however, requires time. I have already (Por, 1975 a) expressed the view that the frequent climatic changes during the Pleistocene resulted in the sea in fluctuating advances and retreats ("pulsations") of species which did not change their ecological requirements. The time–stability dimension necessary for evolution of a new taxon with a different ecological valency simply did not exist during the eventful history of the Pleistocene mediterranean seas. Therefore, I considered (Por, 1975 a) that the Lessepsian influx of species into the Mediterranean became possible since it fitted present interglacial oceanographic conditions in that sea. Under the present conditions, the cold water species complex which flourished during the glacial oceanographic conditions is on the retreat, leaving a vacuum in the subtropical warming-up Levant Sea.

Assuming that drastic climatic fluctuations are bound to continue, the Lessepsian newcomers must be seen as additional actors on the Pleistocene Mediterranean scene. They will expand further west if the climate warms up further, or retreat east and south and eventually disappear if the cold climate returns. Some species of Lessepsian migrant molluscs, especially *Cerithium kochi, Cerithium scabridum, Murex tribulus, Thais carinifera, Pinctada radiata, Brachidontes variabilis, Malleus regula*, and *Paphia textile*, will remain the indicator fossils of the "Lessepsian" facies in the Eastern Mediterranean palaeontology of the future.

In the present human-scale historical period one has to consider that the Lessepsian influence will remain confined to the Eastern Mediterranean with the above-mentioned exceptions of the Aegean and the Adriatic. However, there are changes bound to occur within the coming years. Much has been written concerning the influence of the Aswan High Dam on the process of Lessepsian migration. Some authors have gone as far as to assume that the further depletion of nutrients in the already nutrient-poor Levant Sea (as a result of the cessation of the Nile influx) will in some undefined way add further advantages to the invading Lessepsian migrant species. As noted above, there are indications that after the influx of Lessepsian migrants, better use, or rather additional use is being made of some trophic niches of the Levant coast. However, this is occurring without changing the general trophic framework of the sea.

A considerable contingent of Red Sea organisms are provided with endosymbiotic algae, especially dinoflagellates, and blue–green algae. The coral reefs with their endosymbiontic zooxanthellae are the product of a successful adaptation to a nutrient-poor environment. The zooplankton of the Indian Ocean and the Red Sea—particularly Radiolaria, Foraminifera and even Dinoflagellata—provide their own intracellular nutrient-photosynthetic system through symbiotic zooxanthellae or cyanellae. Taylor (1973) considers that: "the blue-green partners of dynophisoid dinoflagellates are nitrogen fixers as well as carbon fixers" and that: "Such an attribute would be of great value in the Indian Ocean where nitrogen-depleted conditions seem to be much more frequent than phosphorus depletion ..." Reiss (personal communication) studying the globigerinid foraminiferans of the Gulf of Elat also dwells at length on the endosymbiotic nutrition of these planktonic forms. Khmeleva (1967) even thinks that in the Red Sea the role of the radiolaria–zooxanthellae system is more important in terms of primary production than the role of the free-living phytoplankton.

Besides reef and plankton organisms, several other inshore invertebrates, such as the giant clam *Tridacna*, the benthic medusa *Cassiopea*, and perhaps also ascidians and sponges supplement their food by endosymbiontic zooxanthellae. It is important to point out that none of the endosymbiont-harboring Red Sea species appears in the list of Lessepsian migrants. One can only speculate on the reason for this absence; it may be first of all connected with the stenohalinity of these organisms, or with the requirement for high temperatures. It is also possible that the relation between host and alga is secondarily disturbed by a physiological effect of salinity and temperature variations. It is interesting that *Cassiopea andromeda*—a very hardy and euryhaline species, which long ago succeeded in colonizing the Canal—did not advance into the Mediterranean. It will be worthwhile to investigate the relation between normal animal feeding and symbiotic feeding in this medusa under different and extreme environmental conditions.

Until a tropical symbiotrophic species settles in the Levant Sea, there is no reason for speaking of an expanded trophic frame. At the same time, unless such organisms—corals and tropical plankton—settle the Mediterranean, there is no reason to assume that parts of that sea will become a province of the Indo-Pacific realm. As shown above, there is no indication that such a profound qualitative change in the species diversity array of the Lessepsian migration may occur in the near future.

Oren (1970) assumes the possibility of a gradual increase in the total salinity of the Levant Basin following the cessation of the Nile influx. There are as yet no data to indicate if this actually happens, or how considerable this salinity increase might be. It is evident that in the event of a certain increase in the salinity of the Mediterranean, Lessepsian species *already present* will have further competitive advantage. However, there will probably be little enrichment of species. Perhaps only a few species which are presently in the Mediterranean confined to coastal lagoons and the Sirbonic Lagoon, for example, *Acetabularia calyculus*, *Metapenaeus stebbingi*, *Mactra olorina*, and *Crenidens crenidens* may expand their area in the Levant Sea.

It is necessary to close the remarks concerning the future of the Lessepsian migration with a few words on pollution. The Gulf of Suez is increasingly polluted by the nearshore and submarine oilfields such as El Morgan, and Abu Rudeis, etc. Soon the refineries at Suez will work again and the SUMED (Suez–Mediterranean) pipeline will be built. In the navigation of the recently reopened Suez Canal the tankers will occupy an even more preponderant role than before. On the Mediterranean coast are the Ashkelon and Tripoli oil terminals, the existing refineries and oil slicks coming from the Libyan oil fields. Oil pollution which chiefly hits the nearshore organisms, may deeply interfere with the natural evolution of the Lessepsian migration and expansion. Domestic sewage and chemical effluents will also undoubtedly contribute their share. Will the thermal pollution perhaps create optimal environments for the Lessepsian migrants? Until such a beneficial exception is proven, marine pollution in the Eastern Mediterranean should be seen as an indiscriminate menace for all marine life—local or invading.

3.9 Faunal Movements from the Mediterranean into the Red Sea — Anti-Lessepsian Migration

In 1865 Leon Vaillant expected "un mélange des faunes" following the opening of the Suez Canal. In 1902 Tillier was the first to discern the one-sidedness of the migrational phenomena. In later years this became more and more evident. De Lattin (1967, p.46) writes: "It is rather remarkable that for the time being there are scarcely any facts pleading for an opposite migrational direction i.e. from the Mediterranean to the Red Sea" (my translation).

Ben-Tuvia (1966) saw this one-sidedness as a result of the contact between a rich and diverse fish fauna such as that of the Red Sea with an impoverished one in the Mediterranean. Bănărescu and Boşcaiu (1973) see the reason in the greater competitive and expansion capacities of the Red Sea fauna. Recently, Briggs (1974b) asserted that the one-sided faunal movement results from the fact that the Indo-West Pacific species are competitively dominant. Fox (1926), W. Steinitz (1929) and many other authors up to the present think that the reason has to be sought in the fact that the currents from Lake Timsah to the north flow northward during most of the year. However, one cannot dismiss the fact that there is a considerable admixture of Mediterranean species in the Canal itself despite the fact that the migration through the Suez Canal is unidirectional in its results (Por, 1971b). The reasons I saw were on one hand the preadaptation of the Red Sea biota to the conditions of the Suez Canal and of the Levant Basin and on the other hand the competitive "vacuum" in the Eastern Mediterranean.

No such favorable preconditions exist for the Atlantic–Mediterranean fauna, which is already impoverished and near its pessimal conditions in the Levant Basin. To migrate even farther along an increasing salinity and temperature gradient is hardly possible. Moreover, the Red Sea fauna is not impoverished, but a fauna in which many endemic species have evolved in answer to the local conditions. In a later paper (Por, 1975a) I have attempted to place the process of Lessepsian migration into the more general framework of the Pleistocene faunal movements within and into the Mediterranean: the Red Sea migrants thus correspond to a subtropical invasion in response to interglacial subtropical conditions in the Eastern Mediterranean.

Since 1969 when I first used it, the term "Lessepsian Migration" has become synonymous with migration from the Red Sea into the Mediterranean. At that time the most up-to-date listing of the species which migrated the opposite way, i.e. from the Mediterranean to the Red Sea, comprised 16 species (H. Steinitz, 1968), at least half of which were cases of dubious merit. Even for the fish the above-mentioned listing contained only old records (Tillier, 1902), or uncontrolled data (Gohar, 1954) regarding their presence outside the Canal outlet at Suez.

Recently, however, several publications have appeared, which put the exclusive unidirectional nature of the Suez Canal migration in a somewhat changed perspective. Ben-Tuvia (1971b, 1975b) mentioned the capture of two Atlanto-Mediterranean fish in the Gulf of Suez; Fishelson and Rullier (1969) listed Mediterranean Polychaeta which allegedly migrated into the Red Sea; and Ben-Eliahu (1972d) commented favorably on this, giving a few further cases from among the Polychaeta. Schmidt (1971, 1972) while analyzing the Hydrozoa of the Gulf of Elat assumed the possibility of transport from the Mediterranean. Glynn (1972) assumed the same for two species of Isopoda.

A total of 53 species listed below (Table 10) have been at different times and by different authors considered to be Mediterranean immigrants to the Red Sea. Some outright erroneous data such as by Russo (1935), on a whole list of would-be Mediterranean echinoderms, have been deleted from the list. However, as seen below and in the remarks added to the list, most of the cases are of dubious value. The only convincing cases are found among the 13 species of fish, most of which are known only from the Port of Suez or the lagoons of Suez, but their presence in the open Gulf is doubtful. The bulk is formed of 32 species of Polychaeta and 11

Table 10. Anti-Lessepsian migrants mentioned in literature (all known from the Mediterranean)

Taxon and species (species in alphabetical order)	Record of distribution: Suez Canal	Record of distribution: Farthest Indo-Pacific locality	Remarks
Algae			
Acetabularia mediterranea Lamarck	—	Suez (Forskål, 1775)	misidentification? (Lipkin, 1972a)
Porifera			
Euspongia zimocca (O. Schmidt)	—	Sudan (Gohar, 1954)	
Leucandra aspera O. Schmidt	+	Sudan (Gohar, 1954)	
Leucosia nausicae (Schuffner)	—	Gulf of Aqaba (Burton, 1952)	"Belonging more properly to the Mediterranean fauna" (Burton, 1952)
Plecortis simplex F. E. Schulze	—	Sudan (Gohar, 1954)	—
Pseudosuberites mollis Topsent	—	Gulf of Aqaba (Burton, 1952)	"Belonging more properly to the Mediterranean fauna" (Burton, 1952)
Tethya aurantium (Pallas)	—	Gulf of Aqaba (Burton, 1952)	
Hydrozoa			
Corydendrium parasiticum (L.)	Port Taufiq	Aden (Thornley, 1908)	"Certainement par la Mediterranée" (Billard, 1926)
Eudendrium racemosum mucronatum Billard	+	—	Presumed Mediterranean (Billard, 1926)
Helgicirrha schulzei Hartlaub	—	Dahlak (Schmidt, 1972)	"Undoubtedly transported through the Suez Canal" (Schmidt, 1972)
Köllikerina fasciculata (Peron et Lesueur)	—	Arabian Sea (Schmidt, 1972)	
Podocoryne meteoris Thiel	—	Red Sea, widespread (Schmidt, 1972)	
Tubularia larynx Ellis et Solander	("Doubtful": Billard, 1926)	Japan	Also St. Helena!
Tubularia mesembryanthemum Allman	—	Pacific	"may cause some discussion" (Schmidt, 1972)
Ventromma halecioides (Alder)	All the Canal (Billard, 1926)	Amboina (Indonesia)	"around the world" (Schmidt, 1972)
Polychaeta			
Clymene collaris (Claparède)	—	Dahlak (Fishelson and Rullier, 1969)	
Clymene lumbricoides Quatrefages	—	Dahlak (Fishelson and Rullier, 1969)	
Dasychone lucullana (Delle Chiaje)	+ (Potts, 1928)	Dahlak (Fishelson and Rullier, 1969)	"variant of *D. cingulata*" (Laubier, 1966)
Eunice pennata (O.F. Müller)	—	Dahlak (Fishelson and Rullier, 1969)	Also South Africa
Eunice torquata Quatrefages	—	Elat (Fishelson and Rullier, 1969)	South Africa, Angola, Cameroun
Hermodice carunculata (Pallas)	—	Dahlak (Fishelson and Rullier, 1969) Indian Ocean	Senegal, Caribbean

Table 10 (continued)

Taxon and species (species in alphabetical order)	Record of Distribution		Remarks
	Suez Canal	Farthest Indo-Pacific locality	
Lumbrineris coccinea (Renier)	+ (Ben-Eliahu, 1972d)	Dahlak (Fishelson and Rullier, 1969)	
Lysidice ninetta Audouin et Milne Edwards	—	Et Tur (Fishelson and Rullier, 1969) Australia	
Nephthys inermis Ehlers	—	Gulf of Suez (Fauvel, 1933)	
Notomastus profundus Eisig	—	Dahlak (Fishelson and Rullier, 1969)	
Paralacydonia paradoxa Fauvel	—	Gulf of Suez (Fauvel, 1933)	
Petaloproctus terricola Quatrefages	—	Red Sea? (Fishelson and Rullier, 1969)	
Pista cristata (Müller)	—	Dahlak (Fishelson and Rullier, 1969) Indian Ocean	
Placostegus tridentatus (Fabricius)	—	Gulf of Suez (dubious) (Fauvel, 1933)	
Poecilochaetus serpens Allen	—	Ras Muhammad (Fishelson and Rullier, 1969) South Africa, Australia	
Pomatoceros triqueter L.	—	Gulf of Suez (Fauvel, 1933)	
Potamilla stichophthalmus (Grube)	—	Gulf of Suez (Fauvel, 1933)	
Scalisetosus fragilis Claparède	+ (Ben-Eliahu, 1972d)	Red Sea (Hartmann-Schröder, 1960)	
Spinther miniaceus Grube	—	Gulf of Suez (Fauvel, 1933)	"On ship" (Fauvel, 1933)
Terebella lapidaria L.	—	Dahlak (Fishelson and Rullier, 1969) Hawaii	
Sipunculida			
Aspidosiphon mülleri Diesing	—	Suez; Gulf of Tadjourah; Japan	
Phascolion strombi (Montagu)	—	Elat, 1 specimen (Wesenberg-Lund, 1957)	
Crustacea			
Chelura terebrans Philippi	+	Suez (Schellenberg, 1928); South Africa	New Zealand
Cymodoce truncata Leach	+	Ethiopia (Nobili, 1906)	"Rapid and successful colonization of the Red Sea" (Glynn, 1972)
Sphaeroma serratum (Fabricius)	+	—	West Africa
Acari			
Litarachna divergens Walter	Kabret, Gulf of Suez (Soar, 1927)	—	"Hitherto only Mediterranean and Atlantic coast of France" (Soar, 1927)
Mollusca			
Cerastoderma glaucum (Bruguiere)	+	Lagoons of N. Gulf of Suez	= *Cardium edule*

Chama gryphoides Lamarck	Port Taufiq and and Suez (Tomlin, 1927)		
Nassa gibbosula (L.)	+	Gulf of Suez (Tillier and Bavay, 1905)	
Ascidiacea			
Trididemnum sargassicola cereum Giard	Suez (Harant, 1927)	—	
Pisces			
Argyrosoma regius (Asso)	+ (Tortonese, 1964)	—	
Dicentrarchus labrax (L.)	"Suez", rare (Keller, 1882)	El Bilaiyim Lagoon, Gulf of Suez (Paperna, pers. comm.)	
Dicentrarchus punctatus (Bloch)	"Suez", rare (Tillier, 1902)	El Bilaiyim (Ben-Tuvia, 1972)	
Engraulis enchrassicolis (L.)	"d'une mer à l'autre" (Tillier, 1902)	—	"said to be at Suez" (Norman, 1927)
Hippocampus brevirostris Cuvier	Port Taufiq (Norman, 1927)	—	
Lichia amia Cuvier et Valenciennes	Suez lagoons, rare (Tillier, 1902)		
Liza aurata (Risso)	+	El Bilaiyim (Ben-Tuvia, 1972)	= *Mugil auratus*
Mugil cephalus L.	d'une mer à l'autre" (Tillier, 1902)	Sharm esh Sheikh Shura el Manqata, Gulf of Elat (Paperna, pers. comm.)	Breeds in the Gulf of Elat (Paperna, pers. comm.)
Sciaena aquilla Lacepède	Suez wharfs (Tillier, 1902)	Northern Gulf of Suez (Gohar, 1954)	
Serranus cabrilla (L.)	Passing from Mediterranean to Red Sea (Norman, 1927)	Northern Gulf of Suez (Gohar, 1954)	Red Sea (Kossman, 1877; Klunzinger, 1884)
Solea vulgaris Quesnel	— ? —	—	On the fishmarket of Suez (Keller, 1882; Krukenberg, 1888)
Sparus aurata Cuvier et Valenciennes	Exceptional at Suez (Tillier, 1902)	—	Not found in Red Sea (Ben-Tuvia, 1972)
Syngnathus algeriensis Playfair	Kabret (Norman, 1927)	Bay of Suez (Gohar, 1954)	synonym (Tortonese, 1969: doubts Gohar)
Temnodon saltatrix (Cuvier et Valenciennes)	Suez lagoons, rare (Tillier, 1902)	—	
Umbrina cirrosa (L.)	Suez (Paget, 1923)	Probably in Gulf of Suez (Keller, 1882)	

species of Hydroida. There is no decapod crustacean in the list and the three molluscs *Chama gryphoides*, *Cerastoderma glaucum*, and *Nassa gibbosula* are known only from the outlet of the Canal at Port Taufiq and Suez.

I would like to separate the list into different categories according to criteria of decreasing reliability.

1. Two species of estuarine, euryhaline fish, *Dicentrarchus punctatus* and *Liza aurata*, have been found in relative abundance in the El Bilaiyim lagoon some 200 km south of Suez (Ben-Tuvia, 1971b, 1975b). So far, these are the only two cases which can be equated as counterparts of the Lessepsian migrants. The advance of these Mediterranean species through the Canal can be followed, and they built up populations at a considerable distance from the Canal. Several more fish species are caught around Suez: *Sciaena aquilla*, *Umbrina cirrosa*, *Engraulis enchrassicolis*, and *Hippocampus brevirostris*. Gohar (1954) speaks of *S. aquilla* being caught in the Northern Gulf of Suez. The above mentioned molluscs belong to the same category of Mediterranean species which progressed as far south as Port Taufiq, but no further into the Red Sea. *Cerastoderma glaucum*, the bivalve formerly called *Cardium edule*, has also reached the lagoons of Suez, possibly in pre-Lessepsian times. The isopod *Sphaeroma serratum* too is known to have reached Port Taufiq, but this is a euryhaline species which could possibly have accompanied the other pre-Lessepsians.

An interesting case in itself is that of the fish *Serranus cabrilla*, which is frequently caught in the Northern Red Sea. It is the only representative of this Atlantic genus in the Red Sea and therefore it served as the first discussed example of Mediterranean immigration. *S. cabrilla* was, however, found in the Southern Red Sea by Kossman and Rauber in 1874–1875, and in the same period at Kosseir in the Sudanese Red Sea by Klunzinger (1884). These dates are too early even for a very successful migrant; therefore, the presence of *Serranus cabrilla* in the Red Sea ought to be considered as prior to the opening of the Canal.

2. There are several species prone to be south-bound migrants through the Suez Canal. They are known from the Suez Canal, but spread very rapidly and are known from sites fairly remote from the Canal. Such is the isopod *Cymodoce truncata* found all along the Sinai Coasts to the head of the Gulf of Elat (Glynn, 1972). Glynn considers that *Cymodoce richardsoni* reported by Nobili (1906) from the waters of Eritrea in the Southern Red Sea is synonymous with *C. truncata*. In this case the species was present in the Southern Red Sea as early as 1891–1896 when the two Italian expeditions on the "Scilla" took place. Therefore Glynn (1972) mentions *C. truncata* as a case of "rapid and successul colonization of the Red Sea" (p. 297). It was indeed rapid, if the species was not preexistent to the opening of the Suez Canal. If this was not so, than the possibility of passive transport should be held responsible.

Of the many Polychaeta listed as Mediterranean immigrants, only two are known from the Suez Canal: *Scalisetosus fragilis* which has been found as far south as Ghardaqa (Hartmann-Schröder, 1960) and *Lumbrineris coccinea* reported by Fishelson and Rullier (1969) from Musseri in the Southern Red Sea.

There are five species of Polychaeta found by Fauvel (1933) in the Gulf of Suez, but not in the Canal. These are also candidates for an eventual rapid passage through the Canal as adventives on ships or on other fouling organisms.

Fauvel assumes that one of them, *Spinther miniaceus* which has been found in association with a sponge, has been transported, together with its host, by ship.

Two hydroids may also belong to this category: *Eudendrium racemosum mucronatum*, which according to Billard (1926), reached Port Taufiq (the typical variety of the species is also known from Japan). *Corydendrium parasiticum* found by Billard (1926) at Port Taufiq and in the open Gulf of Suez (Billard, 1933) came in his words "certainement par la Mediterranée"; however, Thornley reported it in 1907 from Aden.

All the above-mentioned cases could, therefore, be considered at best as examples of a rapid and passive transport through the canal.

3. There are several species which have their main and known distribution in the Atlantic region; when they are reported for the first time from the Red Sea the presumption is usually expressed that they came through the Suez Canal. In almost every instance we are dealing with taxa and species in which zoogeographical information is still incomplete.

The case of *Aiptasia diaphana*, a Mediterranean actinian, is symptomatic: H. Schmidt (1972) confines himself to the comment: "*A. diaphana* was till now considered an endemic Mediterranean species. In the meantime I found it also in the Gulf of Aqaba. According to Carlgren (1949) this species is present also in the Suez Canal". H. Schmidt did not make further comments in this case. Other authors confronted with a similar situation went into suppositions. In the case of *Clymene lumbricoides* and *Terebella lapidaria*, two polychaetes found at Musseri in the Southern Red Sea, Fishelson and Rullier (1969) wrote: "Comme elle *semble* inconnue dans l'océan Indien, c'est de la Méditerranée qu'elle a *dû* passer en mer Rouge par le canal de Suez" (my italics).

H.-E. Schmidt (1972), speaking of the three hydroids *Podocoryne meteoris* ("widespread in the Red Sea"), *Köllikerina fasciculata* (known also from the Arabian Sea) and *Helgicirrha schulzei* (known only from the Dahlak Archipelago), state that they have been "*undoubtedly* transported through the Suez Canal into the Red Sea" (my italics).

Transport through the Canal in these cases is certainly a possibility which should be taken into account, but further investigations are necessary in order to prove that these species did not reach the Red Sea from the south.

Eventually, they may belong to the same category as the wood-boring amphipod *Chelura terebrans* which was found by Schellenberg (1928) at Suez, and by Omer-Cooper (1927) at Port Said. This is probably a cosmopolitan species, since it extends from the North Atlantic to New Zealand; it could have been brought on ship hulls *also* through the Suez Canal.

4. Many of the list of assumed Mediterranean species are known elsewhere in the Indo–Pacific realm, or from Western tropical Africa. Their newly reported occurrence in the Red Sea is probably due to previous lack of knowledge, since these species could have reached the Red Sea from the southeast or from the tropical Atlantic along the African coasts. They are probably circumtropical species for which the Suez Canal played at most a complementary role in their distribution. As an example, among the polychaetes which possibly migrated from the Mediterranean, Fishelson and Rullier (1969) list *Hermodice carunculata* known from the Mediterranean and Southern Red Sea, and also from the Indian

Ocean, Senegal and the Caribbean Sea; or *Eunice torquata* reported from Lebanon, Israel, Sinai, Elat, and also from South Africa, Angola and Cameroun. The above authors, speaking of *Pista cristata*—also known from the Indian Ocean and the Southern Atlantic and now reported from Musseri—state that: "c'est maintenant *Pista cristata* qui y fait son entrée officielle (en mer Rouge), venant probablement de la Méditerranée".

The case of *Aspidosiphon mülleri*, an Atlantic–Mediterranean and Pacific species of sipunculid, is interesting. The species was found in the Gulf of Aden (Bay of Tadjourah) by Hérubel (1904). Fischer (1914) found in Hérubel's specimens the same arrangement of bidentate and unidentate hooks as in the Mediterranean specimens. He thus wrote: "It can therefore be concluded with certitude that this species migrated from the Mediterranean into the Red Sea". I had the opportunity to check Fischer's type of spinulation on some specimens of *A. mülleri* from the Mediterranean coast of Israel identified by Wesenberg-Lund (1957a) and they did not conform to the above statement (Por, 1973b, 1975c).

Several hydroids listed by H.-E. Schmidt (1971), such as *Tubularia larynx*, *Tubularia mesembryanthemum*, and *Ventromma halecioides* are known from the Pacific as tropicopolitans. Schmidt himself stated, regarding these cases, that "the conclusions ... about this migratory route (i.e. through the Suez Canal) may cause some discussion". It does indeed.

5. Finally, there are several species in which the previous identification has been questioned. There are also a few cases insufficiently documented. For the sake of completeness these cases are also listed.

I dwelt on purpose in such length on the cases which can somewhat crudely be called "Anti-Lessepsian" migrants. The argument is very important for the whole purpose of the present book.

One cannot expect a hermetically unidirectional phenomenon in natural processes involving a wealth of diverse taxa. The most which can be said is that the Lessepsian migration has an "almost exclusively one-way character" (Por et al., 1972). The "Anti-Lessepsian" counterpart is a different process, which cannot be compared to that of Lessepsian migration.

Insofar as Lessepsian migration involves gradual advance and spreading and finally a successful accommodation of the migrants in an alien zoogeographical region, there are differences with the species which went the opposite way:

1. There is a gradual advance of Mediterranean species southwards through the Canal, although to a relatively limited extent. A few Mediterranean species reached the surroundings of Suez, but did not colonize the open Gulf of Suez. The few exceptions, two or more, are euryhaline Mediterranean fish species which have established themselves in or around the lagoons of the Gulf of Suez.

2. All the other cases of Anti-Lessepsian migration are known from animal groups in which passive transport on throughgoing ships is the rule. If the distributional facts in the field, in most cases incomplete, will further support the assumption of transport through the Canal, there is still nothing peculiar about it. Passive transport on ships is a worldwide phenomenon whereby species have crossed the Oceans in every possible sense.

In short, the Anti-Lessepsian migration, in the few cases in which it has occurred, cannot be compared to Lessepsian migration proper. It is merely a spread of some adventive species along a busy shipping lane, perhaps the busiest in the world before 1967, which carried 20% of the world tonnage yearly. One can only wonder how it happened that transport of Mediterranean species into the Red Sea was relatively so restricted.

3.10 The Model of Lessepsian Migration and Other Inter-Oceanic Contacts

In the present, when man's impact on nature is one of the important topics and increasingly investigated subjects, Lessepsian migration has become a textbook case for whoever and for whatever needs to exemplify man's impact on the "pristine" nature. This is not correct. Lessepsian migration is a human artifact only to the extent to which the French diplomat-engineer opened a passageway to marine fauna. The dispersal of the Lessepsian migrants is, however, an entirely natural process, a zoogeographical happening, which encompasses a whole fauna and flora and a whole biotic province in the sea. It has nothing in common with the invasion and explosive success of an *isolated* invading species which reached a new environment fortuitously or as more often, by the aid of man. Lessepsian migration is perhaps the last large-scale zoogeographical event in the very last decades in which the Red Sea and the Mediterranean still preserved a pristine stage. On a par with other contemporary biological phenomena and biotic environments, Lessepsian migration is equally endangered by pollution and by interference of man with the environment.

The phenomena to which Lessepsian migration can be compared are other shipping canals, on one hand, and faunal movements in postglacial mediterranean seas (Por, 1972, 1975a) on the other hand. A selection of these comparable phenomena leads to the following chronological list:

1. The opening of the Black Sea to the Mediterranean, through the Bosphorus $\pm$10,000 BP (Before Present).
2. The contact established between the Baltic Sea and the North Sea about 7000 BP.
3. The artificial and intermittent, historical "Pre-Lessepsian" shipping canal between the Red Sea and the Mediterranean, through the intermedium of the Nile 3200 BP—1300 BP.
4. The Welland Canal between the Atlantic and the Great Lakes, bypassing Niagara falls—1829.
5. The Lessepsian Suez Canal—1869.
6. The Corinth Canal between the Gulf of Corinth (Ionian Sea) and the Gulf of Aigina (Aegean Sea)—1893.
7. The Nord–Ostsee Canal (Kaiser-Wilhelm-Canal) between the Elbe estuary on the North Sea and the Bay of Kiel in the Baltic Sea—1895.
8. The Panama Canal, or Centramerican Canal, between the Gulf of Panama (Pacific Ocean), and the Caribbean Sea (Atlantic Ocean)—1914.
9. The Volga-Don Canal, connecting the Caspian and the Black Sea through these two rivers—1953.

Let us now discuss the eight cases mentioned above, in the order of their increasing similarity with Case No. 5—the Lessepsian Suez Canal.

The Corinth Canal connects two areas of sea inhabited to the extent of our knowledge by a similar marine fauna. However, there must have been differences in the biota of the eastern and western side of the Isthmus of Corinth; these have never been surveyed. An equalization and a mixing most probably occurred through the few kilometers-long Canal. Nothing is known about this either.

The Welland Canal which was opened in 1829, was a final addition to a Canal system started in 1819, connecting the upper reaches of the Hudson River, a tributary of the Atlantic, with the Great Lakes, bypassing the obstacle of the Niagara falls. Aron and Smith (1971) have summed up the changes in the fish fauna of the Great Lakes, following the opening of the Welland Canal. The changes were catastrophic, since two euryhaline estuarine species, the alewife *Alosa pseudoharengus* and the sea-lamprey *Petromyzon marinus* invaded the Lakes. Through food competition (*Alosa*) and direct predation (*Petromyzon*) the invaders nearly exterminated several important commercial fishes of the Great Lakes, such as the Atlantic salmon (*Salmo salar*), the lake trout (*Salvelinus namaycush*), the lake herring (*Leucichthys artedi*), etc. Another species which entered the lakes from the sea, the white perch (*Morone americana*) did not become a pest.

The Welland Canal case is a typical case of invasion into a confined environment (the Great Lakes) of two successful invaders. It may be better compared to the deep disturbances produced by accidental or anthropochoric invaders on oceanic islands like Hawaii or St. Helena. That this is the case is especially evident in the competition between *Alosa* and the autochthonous grazer fishes: according to Aron and Smith (1971) "it became clear that the alewife used only a fraction of the niches occupied by the species that it displaced, thus reducing total fish productivity". Such a situation—extermination and resulting under-exploitation of the trophic basis—cannot occur in an open sea. A closed end of a sea may be ecologically undersaturated with species but this happens not because of a competitive hazard, but because of extreme worsening in the environmental abiotic factors. If this occurs then the impoverishment (or conversely, the penetration of new elements) manifests itself at all the levels of the food web and the whole range of the taxonomic diversity. In the Great Lakes, the invasion seems to have been restricted and impressive only on the level of the fish fauna, and not the whole aquatic flora and fauna.

The Nord–Ostsee Canal did not shorten the way for a faunal interchange between the two seas. The western (North Sea) entrance to the Canal carries the freshwaters of the Elbe and the salinity within the Canal *increases* eastwards until it reaches 15–20‰ in the Bay of Kiel. In fact, this is a paradoxical situation: the North Sea has a normal oceanic salinity of around 35‰, while the natural transition to the Baltic through the Danish Sounds leads through a range of *decreasing* salinities. As far as reported, no Baltic brackish species has used the opportunity to migrate through the Canal westward—a phenomenon which is logically quite possible. The advance of the North Sea–Atlantic fauna into the Baltic indeed occurs through the natural contact of the Sounds (see below) and not through the freshwater barrier of the Nord–Ostsee Canal.

The present Panama Canal, which was also started by Ferdinand de Lesseps, leads through a system of locks, but more important, through two big artificial freshwater lakes—Lake Gatun and Lake Miraflores. According to Abele (1972) Lake Gatun is the most important of the two freshwater barriers along the Canal. The fauna of the Panama Canal (opened in 1913) was first investigated by Hildebrand (1939) and in recent years by Rubinoff (1968, 1970), Rubinoff and Rubinoff (1968), and by Abele (1972). Fauna on both sides was analyzed by Powell (1971) and Alvariño (1974). Discussion of possible passive transport through the Canal can be found in Menzies (1968) and Chesher (1968).

It seems that practically no species was able to cross the Isthmus of Panama through the present Canal. There are old reports that the turtle grass *Thalassia* and the tarpon *Megalops atlanticus* have crossed from the Atlantic into the Pacific. McCosker and Dawson (1975) add to this the pipe fish *Oosthetus lineatus*. However, Voss (1972) states that the appearance of these species on the Pacific side is sporadic and that they did not establish stable populations there. Rubinoff and Rubinoff (1968) reported the presence of a very euryhaline Atlantic goby, *Lophogobius cyprinoides*, in the Miraflores Third Lock, a brackish excavation made in 1943 near the Pacific entrance. McCosker and Dawson (1975) add two more Atlantic species of fish which have established themselves in the Miraflores lock: *Lupinoblennius dispar* and *Hypleurochilus aequipinnis*. Abele (1972) has thoroughly analyzed the decapod fauna of the Panama Canal and assumed that three extremely euryhaline species actually crossed the Isthmus: the prawn *Palaemon pandaliformis*, and two mangrove crabs, *Sesarma rhizophorae* and *S. sulcatum*. The prawn went only as far as the Pacific entrance and the two crabs reached only the Caribbean mouth of the Canal. Alvariño (1974) admits the possibility of interchange between the siphonophorans in the plankton of both sides, and Powell (1971) of that of the bryozoans. Menzies (1968) made an interesting trial and towed a set of intertidal organisms through the Panama Canal from one of the ships: most of them survived. Chesher (1968) drew attention to the possibility of transport in ballast water of ships. Evidence indicates a clear predominance of the movement from the Atlantic to the Pacific. However, Rubinoff and Rubinoff (1968) reported the presence of the Pacific goby *Gobiosoma nudum* and McCosker and Dawson (1975) of the Pacific blenny *Omobranchus punctatus* on the Atlantic side of the Canal. In conclusion: (1) very few extremely euryhaline species or a good swimmer like the tarpon have succeeded in crossing the Isthmus of Panama; (2) these species did not spread into the new sea but remained confined to the canal outlets; (3) passive transport such as fouling or in ship ballast might have eventually occurred. These species are, however, nonspecific adventives, as for example, the crabs *Callinectes sapidus* and *Rithropanopeus harrisi* (Abele, 1972), which have also reached the Eastern Mediterranean and the Black Sea, respectively, in ballast water; and (4) there is no evidence that the zoogeographical phenomena caused by the only 60-year-old Canal are building up towards a qualitative change.

The present-day Panama Canal is an excellent parallel to what was the classical Egyptian-Roman Canal across the Isthmus of Suez. There is, of course, quite a different order of magnitude if we compare the sizes of the two Canals, the number of ships which passed through it, the speed of their passage, and thus, the opportunities for passive transport. However, the Panama Canal has existed hardly more than half a century while the classical Suez waterway has existed for centuries. Comparing it with the Panama freshwater canal it appears reasonable to assume that among the species which are suspected to be "Pre-Lessepsian" migrants there may be some extremely euryhaline species, still confined to the Isthmus and its immediate surroundings, as well as several passively transported species, as in the case of the Panama Canal.

There is a last group of phenomena to which it is interesting to compare Lessepsian migration: these are the faunal movements into the Baltic Sea, the Black Sea and the Caspian Sea, movements which started several thousand years ago and are still going on. In the case of the first two seas these are natural contacts with the adjacent sea; however, these natural contacts are by means of narrow and shallow straits perfectly comparable to the Suez Canal. In the case of the Caspian Sea, the new contact with the Black Sea is through an artificial freshwater passageway. However, here again the Volga–Don Canal is somewhat of an artificial replica of the Pleistocene Manych straits (Zenkevitch, 1963) which also had very close to freshwater salinity values. Some of the faunal movements into these three seas have been reviewed in a previous paper (Por, 1975a). Suffice it to say here that in all these cases we are witnessing the advance of a warm and salt-water faunal element, against a cold-water low salinity element: briefly, an advance of an interglacial fauna. In the case of the landlocked Caspian Sea, the Volga–Don Canal supplied an artificial waterway for a faunal progress for which the potential prerequisites were there.

The same happened in the case of the Suez Canal: it enabled the entrance of a subtropical interglacial faunal element into the Levant Basin which was isolated from any other supply of warmwater fauna. The Lessepsian migration has reached its impressive and unique dimensions only since it is a replica of natural phenomena, an artifact which was unknowingly made in the right place and at the right time.

Much has been written and discussed recently on the ecological impact of a projected sea-level Panama Canal—enabling free intermixing of the Atlantic and Pacific waters. The scientific community was justly alarmed by the possible uncontrolled changes which would occur at both ends because of the penetration of new species. What was especially alarming was the fact that, on one hand, the fisheries (fish and shrimp) on both sides of the Isthmus could have been deleteriously influenced and that current knowledge on the marine biota of both coasts was extremely poor. Previsions ranged, therefore, from catastrophic (Briggs, 1969) to less worried, but by and large, scientists admitted the possibility of harmful effects. These focused almost exclusively on the possible penetration into the Caribbean of the poisonous sea-snake (*Pelamys platurus*) and the coral devouring crown-of-thorns sea star (*Acanthaster planci*) (Rubinoff, 1968; 1970; Topp, 1969, etc.). In 1969, the American National Academy of Sciences appointed a Committee of Ecological Research for the Interoceanic Canal (CERIC). The recommendations of this Committee were summarized by Newman (1972). The need for detailed surveying and investigation of the biota on both sides of the Isthmus has been urged, in order to minimize the potential threats. CERIC considered that "the free mixing of tropical American biotas would enact irreversible changes, not only locally, but, eventually throughout the Indo-Pacific and the tropical Atlantic ... some of these interactions are likely to be quite undesirable and the potential threat as regards a sea-level canal cannot be ignored" (Newman, 1972). Recognizing the priority of economic and other needs in building a big sea-level canal, CERIC asked that "it should be designed in such a way as to minimalize ecological effects". The recommendation asked for establishment of an "antibiotic barrier" through a combination of low salinity and high temperature, on the trajectory of the canal. The Canal Commission appointed by the President of the United States did not take into consideration the fears and recommendations of CERIC among others on the fundamentally wrong affirmation that "significant and rather extensive movement of marine life" had already taken place through the *present* Canal! The Canal Commission went on and decided that "the risk of adverse ecological consequences ... appears to be acceptable" and that no antibiotic barrier should be planned for, since "the need for anything in addition to tidal gates has not been established".

Meanwhile, the construction of the Panama sea-level canal has been shelved for economic and also political reasons. However, sooner or later, the project will come up again, and with it the basic discussion will be reopened.

The whole problem of the effects of a Central American sea-level canal is a matter of prevision, and the truth should be somewhere midway between the pessimists and the "fearless" engineers. It is evident that a faunal interchange will occur and it is inconceivable that anything should be left to chance if prophylaxis is so simple.

What can be learned from the Lessepsian migration concerning any future sea-level canal?

1. The faunal movements will not follow invasional patterns with isolated species playing the main role. The rule is the advance or readjustment of whole biota or faunas.

2. The selection of the migratory biota will be the result of several factors: (a) the hydrographic–edaphic conditions of the canal; (b) the hydrological and edaphic conditions in the vicinity and around the canal outlets; (c) the Pleistocene-historical phase through which the newly connected seas go, i.e. the relation between advancing and retreating biota in the two seas under the present climatic-oceanographical circumstances.

3. Despite the above statement, several taxa will be a priori excluded from migration because of the environment typical to any type of canal. There will be no migration of deep sea forms, of very stenohaline species (since the canal will be either less or more saline than the sea, according to its climatological setting), of holoplanktonic species, of species necessitating clean and wave-swept rocky bottoms.

4. There will be no extermination of the local concurrent species. These will retreat either to the deep or more central parts of their areal, where indirect competition through the abiotic environmental factor will hinder further advance of the newcomers. With respect to commercial fisheries this readjustment may change yields and replace well-priced species with low-commercial-value species.

5. Passive transport on ships or in ballast tanks has no relevance to the problem of migration through canals. This is a process enhanced by human navigation, with or without canals.

6. As every natural environmental phenomenon, canal migration is exposed to severe restriction through danger of pollution. The prevalence of a limited diversity of cosmopolitan pollution-resistant species in the world ocean will definitely be hastened by the existence of interoceanic canals.

7. Therefore, there is an added urgency for investigating interoceanic faunal movements. Many opportunities, such as Lessepsian migration, have already been lost, because of the lack of organized and cooperative research during the different stages of a historical process. However, much like the danger facing irreplaceable climax communities in the sea, the unique opportunities for studying contemporary models of big scale faunal replacements and readjustments should be protected as much as possible from deleterious human interference.

Bibliography

*non vidi

Abele, L. G.: Panamanian Decapod Crustacea: Fauna of the Panama Canal, comparative community structure, and remarks on the social and economic importance of some species. 17e Congr. Intern. Zool. (Monte Carlo, Sept. 1972). Theme No. 3: Les conséquences biologiques des canaux inter-océans. 18 pp. (1972) (mimeo)

Achituv, Y.: Studies on the reproduction and distribution of *Asterina burtoni* Gray and *A. wega* Perrier (Asteroidea) in the Red Sea and Eastern Mediterranean. Israel J. Zool. **18**, 329–342 (1969)

Achituv, Y.: On the distribution and variability of the Indo-Pacific sea star *Asterina wega* (Echinodermata: Asteroidea) in the Mediterranean Sea. Mar. Biol. **18**, 333–336 (1973)

Adams, C. G.: Tertiary Foraminifera in the Tethyan, American, and Indo-Pacific Provinces. In: Aspects of Tethyan Biogeography. Adams, C. G., Ager, D. V. (eds.) Syst. Assoc. Publ. **7**, 195–217 (1967)

*Aharoni, J.: Die Säugetiere Palästinas. Z. Säugetierk. **5** (1930)

Aharoni, J.: From Ashkelon to Rubin. Israel—Land and Nature (Teva VaAretz) **2**, 473—476 (1934) (in Hebrew)

*Aillaud, L.: Les eaux des Lacs Amers et du Lac Timsah (Analyses qualitatives et quantitatives). L'Isthme de Suez **13** (294), 359–365 (1868)

Aleem, A. A.: The recent migration of certain Indo-Pacific algae from the Red Sea into the Mediterranean. The New Phytologist, Oxford, **47**, 88–94 (1948)

Aleem, A. A.: Some new records of marine algae from the Mediterranean Sea with reference to their geographical distribution. Acta Hort. Götheburg. **18**, 275–287, 1948 (1950)

Aleem, A. A.: The occurrence of the sea-grass *Halophila stipulacea* (Forssk.) Asch. on the west coast of Egypt. Bull. Fac. Sci. Univ. Alexandria **4**, 79–84 (1962)

Alvariño, A.: Distribution of Siphonophores in the regions adjacent to the Suez and Panama Canals. 17e Congr. Intern. Zool. (Monte Carlo, Sept. 1972). Theme No. 3: Les conséquences biologiques des canaux inter-océans. 30 pp. (1972) (mimeo) [Also in: Fishery Bull. **72**, 527–546 (1974)]

Anandiades, C.: On the appearance of the fish *Tetrodon spadiceus* (Rich.). Prakt. Hellen. Hydrobiol. Inst. **6**(1), 73–75 (1952)

Andrewartha, H. G., Birch, L. C.: The distribution and abundance of animals. Univ. Chicago Press, 1954

Anonymous: Red Sea and Gulf of Aden Pilot. 10th ed. Hydrographic Dept. Admiralty, London, 1955, 518 pp.

Ariani, A. P., Sarra, V.: Sulla presenza del *Portunus pelagicus* (L.) in acque italiane, con osservazioni sulla morfologia della specie (Crustacea Decapoda). Arch. Botan. Biogeogr. Ital. **45**(4), 186–206 (1969)

Aron, W. I., Smith, S. H.: Ship canals and aquatic ecosystems. Science **174**(4004), 13–20 (1971)

*Ascherson, P.: Die geographische Verbreitung der Seegräser. Petermanns Mitt., 1871, Vol. XVII, 241 pp.

Avnimelech, M. A.: Bibliography of Levant Geology. Jerusalem: Israel Program for Scientific Translations, 1965, x + 192 pp.

Avnimelech, M. A., Boskovitz, V.: Some problems of the present distribution of molluscan shells on the Mediterranean coast of Israel. Bull. Res. Coun. Israel **5B**(2), 178–188 (1955)

*Babič, K.: Prilogi fauni Jadranskago mora (after a review in: Zool. Zentralbl. **18**, 183 (1911))

Băcescu, M.: Citeva animale noi pentru Marea Neagra. Studii si cerc. biol. **2**, 12. Bucharest (1960)

Băcescu, M.: Contribution a la connaissance des Tanaidacés de la Méditerranée Orièntale — 1. Les Apseudidae et Kalliapseudidae des côtes d'Israel. Bull. Res. Coun. Israel (B) **10**(4), 137–170 (1961)

Băcescu, M., Muradian, Z.: Contributions à la connaissance des cumacés de la mer Rouge. Rapp. Comm. Int. Explor. Mer Médit. **22**(4), 83 (1973)

*Bader, H.: Die Bitterseen des Suez Kanals. Petermanns Mitt., 1869, Vol. XV, pp. 471–472

Balss, H.: XIV. Bericht über die Crustacea Decapoda (Natantia und Anomura). Cambridge Expedition to the Suez Canal, 1924. Trans. Zool. Soc. London, **22**, 221–230 (1927)

Balss, H.: The Fishery Grounds near Alexandria. VII. Decapoda. Notes Mem. Fish. Res. Dir. Cairo **15**, 1–45 (1936)

Bănărescu, P.: Principii si probleme de zoogeografie. Bucharest: Ed. Acad. Repub. Soc. Rom. 260 pp. (1970)

Bănărescu, P., Boşcaiu, N.: Biogeografie. Bucharest: Editura Stiintifica, 1973, 302 pp

Bannwarth, D.: Biologische Riff-Untersuchungen im Golf von Suez. Senckenberg Mitt., 1913, pp. 43–50

Barash, Al., Danin, Z.: Mollusca from the stomach of *Sparus auratus* fished in the lagoon of Bardawil. "Argamon". J. Israel Malacol. Soc. **2**(3–4), 97–104 (1971a)

Barash, Al., Danin, Z.: Opisthobranchia (Mollusca) from the Mediterranean waters of Israel. Israel J. Zool. **20**, 151–200 (1971b)

Barash, Al., Danin, Z.: The Indo-Pacific species of Mollusca in the Mediterranean and notes on a collection from the Suez Canal. Israel J. Zool. **21** (3–4), 301–374 (1972)

Barraca, P.S., Thomas, W.D.: Clearing the Suez. Ocean Industry (March 1975)

Baussan, J.: Variations anuelles de niveau le long du Canal de Suez. Rev. Geogr. Phys. Geol. Dyn. **11**, 293–321 (1938)

Bavay, A.: Au sujet du passage d'un mollusque de la Mer Rouge en Méditerranée. Bull. Soc. Zool. France **22**, 199 (1897)

Bavay, A.: Note sur les mollusques du Canal de Suez. Bull. Soc. Zool. France **23**, 161–164 (1898)

Beadnell, H.J.L.: Report on the Geology of the Region of the Red Sea Coast between Quseir and Wadi Ranga. Egypt. Min. Finance, Petrol. Res. Bull. No. 13, Government Press, Cairo, 1–37 (1924)

Beaufort, L.F. de: Zoogeography of the Land and Inland waters. London: Sidgwick and Jackson, 1951, 208 pp

Bebbington, A.: Aplysiid species from Malta with notes on the Mediterranean Aplysiomorpha (Gastropoda, Opisthobranchia). Pubbl. Staz. Zool. Napoli **38**, 25–46 (1970)

Beets, C.: Notes on dredging in the Great Bitter Lake of the Suez Canal. Zool. Meded. Rijksmus. Nat. Hist. Leiden **32**(10), 97–106 (1953)

*Bellan, L.: Contributions à l'étude systematique, bionomique et écologique des Annélides Polychètes de la Méditerranée. Thesis, Fac. Sci. Marseille, 1964

Ben-Eliahu, M.N.: The Polychaeta. In: The Hebrew University—Smithsonian Institution Joint Program: Biota of the Red Sea and the Eastern Mediterranean. Prog. Rep. Appendix Res. Proposal 1970/71. Dept. Zool., Hebrew Univ. Jerusalem, 1970 (mimeo)

Ben-Eliahu, M.N.: A description of *Hydroides steinitzi* n. sp. (Polychaeta: Serpulidae) from the Suez Canal with remarks on the serpulid fauna of the canal. Israel J. Zool. **21**, 77–81 (1972a)

Ben-Eliahu, M.N.: Littoral Polychaeta from Cyprus. Tethys **4**, 85–94 (1972b)

Ben-Eliahu, M.N.: Studies on the migration of the Polychaeta through the Suez Canal. 17e Congr. Intern. Zool. (Monte Carlo, 25–30 Sept. 1972). Theme No. 3: Les conséquences biologiques des canaux inter-océans. 8 pp (1972c) (mimeo)

Ben-Eliahu, M.N.: Polychaeta Errantia of the Suez Canal. Israel J. Zool. **21** (3–4), 189–237 (1972d)

Benson, R.H.: An Ostracodal view of the Messinian Salinity Crisis. In: Messinian Events in the Mediterranean. Amsterdam: Akad. Wetenschap., 1973, pp. 235–242

Benson, R.H.: The Symposium on the Biodynamic Effects of the Messinian Salinity Crisis. Serv. Géol. Tunisie (in press a)

Benson, R. H.: Testing the Messinian Salinity Crisis biodynamically: an introduction. In: Benson, R. H. (ed.). The Biodynamic Effects of the Messinian Salinity Crisis. Amsterdam: Elsevier (in press b)

Benson, R. H., Ruggieri, G.: The end of the Miocene, a time of crisis in the Tethys-Mediterranean history. Ann. Geol. Survey Egypt. **4**, 237–250 (1974)

Benson, R. H., Sylvester-Bradley, P. C.: Deep sea Ostracodes and the transformation of ocean to sea in the Tethys. In: Paléoécologie, Ostracodes. Oertli, H. J. (ed.) Bull. Centre Rech. Pau—SNPA Suppl. **5**, 63—91 (1971)

Ben-Tuvia, A.: Mediterranean fishes of Israel. Bull. Sea Fish. Res. Stn. Haifa (8) 1–40 (1953)

Ben-Tuvia, A.: Collection of fishes from Cyprus. Bull. Res. Coun. Israel **11B**(3), 132–145 (1962)

Ben-Tuvia, A.: Systematics and ecology of Indo-Pacific fishes recently established in the Eastern Mediterranean. Proc. 16th Intern. Cong. Zool. (Washington, D.C., Aug. 1963) **1**, 115 (1963) (Abstract)

Ben-Tuvia, A.: Two siganid fishes of Red Sea origin in the Eastern Mediterranean. Bull. Sea Fish. Res. Stn., Haifa (37), 3–9 (1964)

Ben-Tuvia, A.: Red Sea fishes recently found in the Mediterranean. Copeia **2**, 254–275 (1966)

Ben-Tuvia, A.: Revised list of the Mediterranean fishes of Israel. Israel J. Zool. **20**, 1–39 (1971a)

Ben-Tuvia, A.: On the occurrence of a Mediterranean serranid fish *Dicentrarchus punctatus* in the Gulf of Suez. Copeia **4**, 741–743 (1971b)

Ben-Tuvia, A.: Immigration of fishes through the Suez Canal. 17e Congr. Intern. Zool. (Monte Carlo, Sept. 1972). No. 3. Les conséquences biologiques des canaux inter-océans. 8 pp. (1972) (mimeo)

Ben-Tuvia, A.: Man-made changes in the Eastern Mediterranean sea and their effect on the fishery resources. Mar. Biol. **19**(3), 197–203 (1973)

Ben-Tuvia, A.: Mugilid fishes of the Red Sea with a key to the Mediterranean and Red Sea species. Bamidgeh **27**(1), 14–20 (1975a)

Ben-Tuvia, A.: Comparison of the fish-fauna in the Bardawil lagoon and the Bitter Lakes. Rapp. Comm. Int. Explor. Mer Médit. **23**(3), 125–126 (1975b)

Ben-Tuvia, A.: Fish collections from the Eastern Mediterranean, the Red Sea and inland waters of Israel. Dept. Zool., Zool. Mus., Hebrew Univ. Jerusalem, 1976, 32 pp.

Ben-Tuvia, A., Herman, Z.: The biology of the fishes of Bardawil. Fish. Fishbreed. Israel, **7**, 38–53 (1972) (in Hebrew)

Ben-Tuvia, A., Herman, Z.: Report on the fisheries research in the Bardawil Lagoon. Israel Oceanogr. Limnol. Res. Ltd. Report, February 1973, 30 pp. (mimeo) (in Hebrew)

Ben-Tuvia, A., Lourie, A.: A Red Sea grouper *Epinephelus tauvina* caught on the Mediterranean coast of Israel. Israel J. Zool. **18**, 245–247 (1969)

Ben-Yami, M., Glaser, T.: The invasion of *Suarida undosquamis* (Richardson) into the Levant Basin—An example of biological effect of interoceanic canals. Fishery Bull. **72**(2), 359–373 (1974)

Berdugo, V.: Sur la présence dans la Méditerranée orientale de 2 espèces du genre *Calanopia* (Copepoda: Calanoida). Rapp. Comm. Int. Explor. Mer Médit. **19**(3), 445–446 (1968)

Berdugo, V.: On the occurrence of *Acartia centrura* (Copepoda: Calanoida) in the neritic waters of the Eastern Mediterranean. Rapp. Comm. Int. Explor. Mer Médit. **22**(9), 85–86 (1974)

Berdugo, V., Kimor, B.: Considerations on the distribution of pelagic Copepoda in the Eastern Mediterranean. Rapp. Comm. Int. Explor. Mer Medit. **19**(3), 447–448 (1968)

Berggren, W. A.: Micropaleontologic investigations of Red Sea cores—summation and synthesis of results. In: Hot Brines and Recent Heavy Metal Deposits in the Red Sea. Degens, E. T., Ross, D. A. (eds.) Berlin-Heidelberg-New York: Springer, 1969, pp. 329–335

Berggren, W. A., Boersma, A.: Late Pleistocene and Holocene Planktonic Foraminifera from the Red Sea. In: Hot Brines and Recent Heavy Metal Deposits in the Red Sea. Degens, E. T., Ross, D. A. (eds.) Berlin-Heidelberg-New York: Springer, 1969, pp. 282–298

Bertin, L.: Les clupeiformes du Canal de Suez, comparés a ceux de la Mer Rouge et de la Méditerranée. Bull. Mus. Hist. Natur. Paris (Ser. 2) **15**(6), 386–391 (1943)

Billard, A.: IV. Rapport sur les Hydroïdes. Cambridge Expedition to the Suez Canal, 1924. Trans. Zool. Soc. London, **22**, 85–104 (1926)

Billard, A.: Les Hydroïdes des Golfes de Suez et d'Akaba. Mission Robert Ph. Dollfus en Egypte (Decembre 1927–Mars 1929). Mém. Inst. Egypte, **21**, 1–30 (1933)

Blanc, J. J.: The sedimentary geology of the Mediterranean Sea. Oceanogr. Mar. Biol. Ann. Rev. Barnes, H. (ed.). **6**, 377–454 (1969)

Bombace, G.: Sul rinvenimento di alcuni esemplari di *Pinctada radiata* (Leach) nelle acque del Canale di Sicilia (Mollusca, Bivalvia). Natura (Milano) **58**(4), 298–304 (1967)

Bonalti, L.: North Mediterranean climate during the last Würm glaciation. Nature (London) **209**, 984 (1966)

*Bourdon, C.: Anciens canaux et anciens sites et ports de Suez. Mém. Soc. Roy. Géog. Egypte. (1925)

Briggs, J. C.: Panama's sea-level canal. Science, **162**, 511–513 (1968)

Briggs, J. C.: The sea-level Panama Canal: potential biological catastrophe. Bioscience **19**(1), 44–47 (1969)

Briggs, J. C.: Operation of zoogeographic barriers. Syst. Zool., **23**, 248–256 (1974a)

Briggs, J. C.: Marine Zoogeography. New York: McGraw-Hill, 1974b, 475 pp.

Broch, Hj.: VII. Report on the Crustacea Cirripedia. Cambridge Expedition to the Suez Canal, 1924. Trans. Zool. Soc. London **22**, 133–138 (1927)

Browne, E. T.: V. Report on the Medusae. Cambridge Expedition to the Suez Canal, 1924. Trans. Zool. Soc. London **22**, 105–115 (1926)

Brunelli, G., Bini, G.: Sulla immigrazione di una specie di "*Theuthis*" dal Mar Rosso al Mare Egeo. R. C. Acad. Lincei, Roma **19**(4), 255–260 (1934)

Bruun, F. A.: *Parexocoetus*, a Red Sea flying fish in the Mediterranean. Nature (London) **136**, 553 (1935)

Burfield, S. T.: XXI. Report on the Chaetognatha. Cambridge Expedition to the Suez Canal, 1924. Trans. Zool. Soc. London **22**, 355–357 (1927)

Burkenroad, M. D.: Littoral Penaeidea, chiefly from the Bingham Oceanographic Collection. With a revision of *Penaeopsis*, etc. Bull. Bingham Oceanogr. Coll. **4**(7), 1–109 (1934)

Burton, M.: III. Report on the sponges. Cambridge Expedition to the Suez Canal, 1924. Trans. Zool. Soc. London **22**, 71–83 (1926)

Burton, M.: The Fishery Grounds near Alexandria. IX. Sponges. Notes Mem. Fish. Res. Dir. Cairo (17), 1–28 (1936)

Burton, M.: The 'Manihine' Expedition to the Gulf of Aqaba. III. Sponges. Bull. Brit. Mus. (Nat. Hist.) Zool. **1**(8), 163–174 (1952)

Butzer, K. W.: Studien zum vor- und frühgeschichtlichen Landschaftwandel der Sahara. III. Die Naturlandschaft Ägyptens während der Vorgeschichte und der Dynastischen Zeit. Ak. Wiss. Litt. Math. Nat. Klasse 2, 51–122 (1951)

Butzer, K. W.: Contributions to the Pleistocene Geology of the Nile Valley. Erdkunde **13**, 46–67 (1959)

Butzer, K. W.: Mediterranean area: Quaternary history. In: The Encyclopedia of Oceanography. Fairbridge, R. W. (ed.). New York: Van Nostrand, Reinhold, 1966, pp. 485–490

Butzer, K. W.: Environment and Archeology—an Ecological Approach to Prehistory. London: Methuen, 1971, 703 pp.

Calman, W. T.: XIII. Report on the Crustacea Decapoda (Brachyura). Cambridge Expedition to the Suez Canal, 1924. Trans. Zool. Soc. London **22**, 211–219 (1927a)

Calman, W. T.: XXVII. Report on the Phyllocarida, Cumacea, and Stomatopoda. Cambridge Expedition to the Suez Canal, 1924. Trans. Zool. Soc. London **22**, 399–401 (1927b)

Calman, W. T.: XXVIII. Report on the Pycnogonida. Cambridge Expedition to the Suez Canal, 1924. Trans. Zool. Soc. London **22**, 403–410 (1927c)

Capper, J.: Observations on the Passage to India through Egypt and across the Great Desert. London: 1783

Carlgren, O.: XXXI. Report on the Actiniaria and Ceriantharia. Cambridge Expedition to the Suez Canal, 1924. Trans. Zool. Soc. London **22**, 443–445 (1927)

Carmin, J.: Algae of Palestinian shores. Bull. Inst. Océanogr. Monaco (653), 1–7 (1934)

Carmin, J.: Mollusca of Palestinian Shores, first report. Bull. Indep. Biol. Lab. (Kefar-Malal), **4** No. 4, (38, 39) 1–9 (1946)

Carus, J. V.: Prodromus faunae Mediterraneae. Stuttgart: Schweizerbart 1889–1893, Vols. I and II. 854 pp.

Casson, L.: The Ancient Mariners—Seafarers and Sea Fighters of the Mediterranean in Ancient Times. New York: Mac Millan, 1959, 286 pp.

Chabanaud, P.: Poissons recueillis dans le grand Lac Amer (Isthme de Suez) par M. le Professeur A. Gruvel en 1932. Bull. Mus. Hist. Natur. Paris (Ser. 2) **4**(7), 822–835 (1932)

Chabanaud, P.: Sur divers poissons de la Mer Rouge et du Canal de Suez. Description de deux éspèces nouvelles. Bull. Inst. Océanogr. Monaco **30**(627), 1–12 (1933a)

Chabanaud, P.: Contribution a l'étude de la faune ichthyologique du Canal de Suez. Bull. Soc. Zool. France **58**, 287–292 (1933b)

Chabanaud, P.: Poissons recueillis dans le lac Timsah (Isthme de Suez) par M. le Professeur A. Gruvel, en 1933. Bull. Mus. Hist. Natur. Paris (Ser. 2) **6**(1), 156–160 (1934)

Chakroun-Ktari, F.: Captures d'animaux rares en Tunisie. Bull. Inst. Nat. Sci. Techn. Océanogr. Pêche Salambô N.S. **1**(2), 75–79 (1966)

Chesher, R. H.: Transport of marine plankton through the Panama Canal. Limnol. Oceanogr. **13**(2), 387–388 (1968)

Christiaens, J.: Présence de la patelle, *Cellana rota* Gmelin en Méditerranée Orientale. Bull. Mus. Nat. Hist. Natur. Paris (Ser. 2) **38**, 903 (1967)

Codreanu, R.: Clasificarea evolutiva a bopirienilor, izopode parazite ale crustaceilor decapode şi importanta lor biologica generala. Stud. şi Cercetari Biol. Ser. Zool. **10**(3), 203–211 (1967)

Coleman, R. G.: Geologic Background of the Red Sea. In: Initial Reports of DSDP. 23. Whitmarsh, R. B., Weser, O. E., Ross, D. A. (eds.). U.S. Govt. Printing Press, 1974, pp. 813–819

Collette, B. B.: *Rastrelliger kanagurta*, another Red Sea immigrant into the Mediterranean Sea, with a key to Mediterranean species of Scombridae. Bull. Sea Fish. Res. Stn. Haifa (54), 3–6 (1970)

Cox, L. R.: Notes on the Post-Miocene Ostreidae and Pectinidae of the Red Sea Region, with remarks on the geological significance of their distribution. Proc. Malac. Soc., **18**, 165–209 (1929)

Czoering Freih, V.: Salz aus dem großen Bittersee im Suezkanal. Verh. Kais. Konig. Geol. Reichsanst. Wien **10**, 222–223 (1869)

Dammerman, K. W.: The Fauna of Krakatau 1883–1933. Koninkl. Ned. Acad. Wetenschap. (3), 44 (1948)

Dansereau, P.: Biogeography, an Ecological Perspective. New York: Ronald Press, 1957, 394 pp.

Darlington, P. J.: Zoogeography, The Geographical Distribution of Animals. London-New York: Wiley, 1957, 675 pp.

Darlington, P. J.: Area, Climate, and Evolution. Evolution **13**, 488–510 (1959)

*Dautzenberg, Ph.: Mollusques marins recueillis sur les Côtes de Tunisie et d'Algerie. Mém. Soc. Zool. France 363, 373 (1895)

Dean, L. M. I.: XXXVII. Report on the Alcyonaria. Cambridge Expedition to the Suez Canal, 1924. Trans. Zool. Soc. London, **22**, 707–712 (1929)

Demetropoulos, A.: Marine molluscs of Cyprus. Part A. Placophora, Gastropoda, Scaphoda, Cephalopoda. Fisheries Bull. Min. Agric. Nat. Res. Cyprus, No. 2 (1969)

Demetropoulos, A.: Marine molluscs of Cyprus. Part B.—Bivalvia (Lamellibranchiata). Fisheries Bull. Min. Agric. Nat. Res. Cyprus, Nos. 3, 5 (1971)

Demetropoulos, A., Neocleous, D.: The fishes and crustaceans of Cyprus. Fisheries Bull. Min. Agric. Nat. Res. Cyprus, No. 1, 3–21 (1969)

Deuser, W. G., Degens, E. T.: O^{18}/O^{16} and C^{13}/C^{12} ratios of fossils from the hot-brine deep areas of the Central Red Sea. In: Hot Brines and Recent Heavy Metal Deposits in the Red Sea. Degens, E. T., Ross, D. A. (eds.). Berlin-Heidelberg-New York: Springer, 1969, pp. 336–374

Dollfus, R. Ph.: Mission Robert Ph. Dollfus en Egypte, X. Stomatopoda. Mém. Inst. Egypte **37**, 185–236 (1938)

Dowidar, N. M.: Distribution and ecology of *Ceratium egyptiacum* Halim and its validity as indicator of the current regime in the Suez Canal. J. Mar. Biol. Ass. India **15**(1), 335–344 (1973)

Dowidar, N. M.: Tintinnids from the Suez Canal. Rapp. Comm. Int. Explor. Mer. Médit. **22**(9), 123–124 (1974)

Durand-Claye, L.: Essai des vingt-un échantillons d'eau salée du Canal maritime de Suez. Annal. Chim. Phys. (Ser. 5) **3**, 188 (1874)

Eales, N. B.: On the migration of Tectibranch Molluscs from the Red Sea to the Eastern Mediterranean. Proc. Malac. Soc. London **39**, 217–220 (1970)

Eales, N. B., Engel, H.: The genus *Bursatella* de Blainville. Proc. Malac. Soc. London **21**, 297–303 (1935)

Ehrlich, A.: The diatoms from the surface sediments of the Bardawil Lagoon (Northern Sinai)—Paleoecological significance. In: Simonsen, R. (ed.). Proc. 3rd Symp. Recent and Fossil Marine Diatoms (Kiel, Sept. 9–13, 1974). Beiheft 53 zur Nova Hedwigia, 253–277 (1975a)

Ehrlich, A.: Les Diatomées benthiques et épiphytes de la lagune de Bardawil (Sinai septentrional). Rapp. Comm. Int. Explor. Mer Médit. **23**(3), 121–123 (1975b)

Eitan, G.: Additions to the Bryozoan Fauna of the Suez Canal. Israel J. Zool. **21** (3–4), 377–384 (1972)

Eitan, G.: Studies on Post-Larval Development in Marine Bryozoa (metamorphosis, colony formation and regeneration). Ph. D. thesis. Hebrew Univ. Jerusalem, 1974, 71 pp. (in Hebrew with English summary)

Ekman, S.: Begründung einer statistischen Methode in der regionalen Tiergeographie, nebst einer Analyse der paläarktischen Steppen und Wüstenfauna. Nova Acta. Reg. Soc. Sci. Upsaliensis, **4**(12), 117 pp. (1940)

Ekman, S.: Zoogeography of the Sea. London: Sidgwick and Jackson, 1967, XIV + 417 pp.

El-Sabh, M. I.: Effect of the Aswan High Dam on the distribution of salinity in the Suez Canal. Nature (London) **218**(5143), 758–760 (1968)

El-Sabh, M. I.: Seasonal hydrographic variations in the Suez Canal after the completion of the Aswan High Dam. Kiel. Meeresforsch. **25**(1), 1–18 (1969)

Emery, K. O., Bentor, Y. K.: The continental shelf of Israel. Bull. Geol. Surv. Israel **26**, 25–40 (1960) [Also in: Bull. Sea Fish. Res. Stn., Haifa (28) (1960)]

Emery, K. O., Hunt, J. M., Hays, E. E.: Summary of hot brines. In: Hot Brines and Recent Heavy Metal Deposits in the Red Sea. Degens, E. T., Ross, D. A. (eds.). Berlin-Heidelberg-New York: Springer, 1969

Emiliani, C.: Pleistocene temperature variations in the Mediterranean. Quaternaria **2**, 87–98 (1955)

Emiliani, C.: Cenozoic climatic changes as indicated by the stratigraphy and chronology of Deep Sea cores of Globigerina-ooze facies. Ann. N. Y. Acad. Sci. **95**, 521–536 (1961)

Engel, H., Eeken, C. J. van: Red Sea Opisthobranchia from the coast of Israel and Sinai. Bull. Sea Fish. Res. Stn., Haifa (30), 15–34 (1962)

Erazi, R. A.: *Leiognathus mediterraneus* nov. sp. C. R. Soc. Turq. Sci. Phys. **10**(10), 49–53 (1943)

Evans, G., Murray, J. W., Biggs, H. E. J., Bate, R., Bush, P. R.: The oceanography, ecology, sedimentology, and geomorphology of parts of the Trucial Coast Barrier Island Complex. In: The Persian Gulf. Purser, B. H. (ed.). Berlin-Heidelberg-New York: Springer, 1973, pp. 233–277

Fairbridge, R. W.: Eustatic changes in sea level. In: Physics and Chemistry of the Earth. Ahrens, L. H. (ed.) New York: Pergamon Press, 1961, pp. 99–185

Fairbridge, R. W.: Mean sea level changes, long-term-eustatic and other. In: The Encyclopedia of Oceanography. Fairbridge, R. W. (ed.). New York: Van Nostrand Reinhold, 1966, pp. 479–485

Fairbridge, R. W.: Quaternary sedimentation in the Mediterranean region controlled by tectonics, paleoclimates and sea level. In: The Mediterranean Sea: a Natural Sedimentation Laboratory. Stanley, D. J. (ed.). London: Hutchinson and Ross, 1972, pp. 99–113

*Faouzi, H.: Régime des courants dans le Canal de Suez. Hydrobiol. Fisheries Dir. Cairo (23) (1936)

Faouzi, H.: Le Canal des Suez, voie d'échanges biologiques entre la Méditerranée et la Mer Rouge. Ann. centre Univ. Médit. **5**, 23–30 (1951–1952)

Faurot, L.: Observation au sujet des mollusques testacés recueillies par MM. Tillier et Bavay dans le Canal de Suez. Bull. Soc. Zool. France **31**, 42–45 (1906)

Fauvel, P.: XXIX. Rapport sur les Annélides polychètes errantes. Cambridge Expedition to the Suez Canal, 1924. Trans. Zool. Soc. London **22**, 411–439 (1927)

Fauvel, P.: Annélides polychètes. In: Mission Robert Ph. Dollfus en Egypte. Mém. Inst. Egypte **21**, 31–83 (1933)

Fauvel, P.: Annélides Polychètes de Rovigno d'Istria. Thalassia **1**, 1–78 (1934)

Fauvel, P.: Les fonds de pêche près d'Alexandrie, XI. Annélides polychètes. Notes Mem. Fish. Res. Dir., Cairo **19**, 1–60 (1937)

Fauvel, P.: Annélides Polychètes de Haute-Adriatique. Thalassia **4**(1), 1–24 (1940)

Fauvel, P.: Contribution à la faune des Annélides polychètes des côtes d'Israël. I. Bull. Sea Fish. Res. Stn. Haifa (10), 1–12 (1955)

Fauvel, P.: Contribution à la faune des Annélides polychètes des côtes d'Israël. II. Bull. Res. Coun. Israel **6B**(3–4), 213–219 (1957)

*Feldmann, J.: Note sur quelques algues marines de Tunisie. Stn. Océanogr. Salambô, Tunisie, Notes, no. 24 (1931)

Feldmann, J., Feldmann, G.: Présence du *Lophocladia Lallemandi* (Mont.) Schmitz, aux environs d'Alger. Bull. Soc. Hist. Nat. Afrique du Nord **29**, 479 (1938)

Feldmann, J., Feldmann, G.: Additions à la flore des Algues marines de l'Algerie. Bull. Soc. Hist. Nat. Afrique du Nord, Fasc 4(30), 80–91 (1947)

Felsenburg, T.: *Brachidontes variabilis* (Bivalvia, Mytilidae)—as a successful coloniser. Preliminary Report. M.Sc. thesis, Hebrew Univ. Jerusalem, 1974 (in Hebrew)

Fenaux, R.: Sur les appendiculaires de la Mer Méditerranée orientale. Bull. Mus. Nat. Hist. Natur. Paris, (2), **42**, 1208–1211 (1971)

*Fischer, P. H.: Notes sur les faunes conchyliologiques des deux rivages de l'Isthme de Suez. J. Conchyl. **12**, 241–248 (1865)

Fischer, P. H.: Sur la faune conchyliologique marine des baies de Suez et de l'Akabah. J. Conchyl. **18**, 161–179 (1870)

Fischer, P. H.: Sur la faune conchyliologique marine de la baie de Suez. J. Conchyl. **19**, 209–226 (1871)

Fischer, W.: Weitere Mitteilungen über die Gephyreen des Naturhistorischen Museums zu Hamburg. Jb. Hamburg. Wiss. Anst. 31, 1–28 (1914)

Fishelson, L., Rullier, F.: Quelques Annélides polychètes de la mer Rouge. Israel J. Zool., **18**, 49–117 (1969)

*Fonselius, S. H.: Unpublished manuscript. In: Meddelande från Havsfiske-laboratoriet Lysekil, No. 19, Goteborg, 1966

*Fonselius, S. H.: Analysis of trace metals in the Mediterranean and the Indian Ocean. P. V. Ass. Océanogr. phys. géod. intern. (10), 219–220 (1967)

Forest, J., Guinot, D.: Sur une collection de crustacés décapodes des côtes d'Israël. Bull. Sea Fish. Res. Stn. Haifa (15), 4–16 (1958)

Forskål, P.: Descriptiones animalium, avium, amphibiorum, piscium, insectorum, vermium, quae in itinere orientali observavit Petrus Forskal ... (Post mortem auctoris edidit Carsten Niebuhr. [Adjuncta est materia medica Kahirina atque tabula maris Rubri geographica] Hauniae (Copenhagen), 1775, 1–19, i-xxxii, 1–164

Forti, A.: Nuove segnalazioni di Alghe passivamente transportate a traverso al canale di Suez, poi naturalizzate. Nuovo. G. Bot. Ital. n.s., **34**(2), 1443–1451 (1928)

Fowler, H. W., Steinitz, H.: Fishes from Cyprus, Iran, Iraq, Israel and Oman. Bull. Res. Coun. Israel **5B**(3–4), 260–292 (1956)

Fox, H. M.: The migration of a Red Sea crab through the Suez Canal. Nature (London) **113**, 714–715 (1924)

Fox, H. M.: The Biology of the Suez Canal. Nature (London) **115**(2886), 262–263 (1925)

Fox, H. M.: I. General part. Zoological Results Cambridge Expedition to the Suez Canal, 1924, Trans. Zool. Soc. London **22**, 1–64 (1926)

Fox, H. M.: XXXIX. Summary of results. Zoological Results Cambridge Expedition to the Suez Canal, 1924. Trans. Zool. Soc. London **22**, 843–863 (1929)

Fraser-Brunner, A.: Notes on the Plectognath Fishes. III. On *Monacanthus setifer* Bennett and related species, with a key to the genus *Stephanolepis* and description of four new species. Ann. Mag. Nat. Hist. Ser. 11, 5(30), 518–535 (1940)

Fredj, G.: Stockage et exploitation des données en écologie marine. A—Un fichier sur ordinateur des Invertébres macrobenthiques. Mém. Inst. Oceanogr. Monaco, **4**, 61 pp. (1972)

Fredj, G.: Stockage et exploitation des données en écologie marine. C — Considérations biogéographiques sur le peuplement benthique de la Méditerranée. Mém. Inst. Oceanogr. Monaco, **7**, 88 pp. (1974)

Friedman, G. M.: Geology and geochemistry of reefs, carbonate sediments, and waters, Gulf of Aqaba (Elat), Red Sea. J. Sediment. Petrol. **38**(3), 895–919 (1968)

Fritsch, C.: Über die Auffindung einer marinen Hydrocharidae im Mittelmeer. Verh. Zool.-Bot. Ges. Wien **45**, 104–106 (1895)

Fröiland, O.: Fishes of the family Scorpaenidae from Cyprus, including three new records. Bull. Sea Fish. Res. Stn., Haifa (59), 5–16 (1972)

Fuchs, Th.: Die geologische Beschaffenheit der Landenge von Suez. Denkschr. k. Akad. Wissenschaft. (Math. Nat. Kl.), Wien, **38**, 25–42 (1878)

*Fuchs, Th.: Über die geologische Beschaffenheit der Landenge von Suez und den Amurliman im nordjapanischen Meere. Verh. Geol. Reichsanstalt Wien, 178–181 (1881)

Furnestin, M. L.: Sur quelques chaetognathes d'Israel. Bull. Res. Coun. Israel **2**(4), 411–414 (1953) [Also: Bull. Sea Fish. Res. Stn. (6), 411–414 (1953)]

Gardiner, J. S.: The Biology of the Suez Canal. Nature (London) **114**(2865), 520–521 (1924)

*Gauthier, V.: Sur les Echinides qui vivent aux environs de Marseille. Compt. Rend. Acad. Sci. (Paris) **79**, 403 (1874)

Geldiay, R., Kocatas, A.: A report on the occurrence of Penaeidae (Decapoda Crustacea) along the coast of Turkey from Eastern Mediterranean to the vicinity of Izmir, as a result of migration and its factors. 17e Congr. Intern. Zool. (Monte Carlo, Sept., 1972). Theme No. 3. Les conséquences biologiques des canaux inter-océans. (1972) 7 pp. (mimeo)

George, C. J., Athanassiou, V. A.: On the occurrence of *Scomberomorus commersoni* (Lacépède) in St. George Bay, Lebanon. Doriana, Ann. Mus. Stor. Nat. Genova **4** (157), 237–294 (1965)

George, C. J., Athanassiou, V. A.: Additions to the check list of the fishes of the coastal waters of Lebanon. Misc. Pap. Nat. Sci. Am. Univ. Beirut, **5**, 6–8 (1966)

George, C. J., Athanassiou, V. A., Boulos, I.: The fishes of the coastal waters of Lebanon. Misc. Pap. Nat. Sci. Am. Univ. Beirut **4**, 1–27 (1964)

Geronimo, I.: Prima segnalazione nelle coste italiane di *Brachidontes variabilis* Krauss. Boll. Sedute Accad. Gioenia Sci. Nat. Catania **10**, 847–852 (1971)

Ghazzawi, F. M.: Plankton of the Egyptian waters. A Study of the Suez Canal Plankton. (A) The Phytoplankton. Notes Mém. Fish. Res. Dir. Cairo (24), 1–83 (1936)

Ghazzawi, F. M.: Plankton of the Egyptian waters. Two Cladocera from the plankton. Notes Mém. Fish. Res. Dir. Cairo (31), 1–2 (1938)

Ghisotti, F.: Molluschi del genere *Quoyula* Iredale 1912 nel Mediterraneo. Conchiglie, Unione Malacol. Ital., Anno IV, 11–12, 182–186 (1968)

Giesbrecht, W.: Systematik und Faunistik der pelagischen Copepoden des Golfes von Neapel und der angrenzenden Meeresteile. Fauna und Flora des Golfes von Neapel. Berlin: 1892

Giesbrecht, W.: Über pelagische Copepoden des Rothen Meeres. Zool. Jb. Syst. **9**, 315–328 (1896)

Gilat, E.: The macrobenthonic invertebrate communities on the Mediterranean continental shelf of Israel. IAEA Radioactivity in the Sea. Publ. No. 8. Bull. Inst. Océanogr. Monaco, **62**, 1–46 (1964)

Gilat, E.: The macrobenthic communities of the level bottom in the Eastern Mediterranean. In: The Hebrew University—Smithsonian Institution Joint Program: Biota of the Red Sea and the Eastern Mediterranean. Interim Report. Dept. Zool., Hebrew Univ. Jerusalem, 1969, pp. 82–89 + tables (mimeo)

Gitay, A.: A review of *Augeneriella* (Polychaeta: Sabellidae) and a new species from northern Sinai. Israel J. Zool. **19**, 105–109 (1970)

Glynn, P. I.: Isopoda of the Suez Canal. Israel J. Zool. **21**(3–4), 275–300 (1972)

Godeaux, J.: A contribution to the knowledge of the Thaliacean faunas of the Eastern Mediterranean Sea and the Red Sea. Israel J. Zool. **22**, 39–50 (1973)

Godeaux, J.: Thaliacés récoltés au large des côtes egyptiennes de la Méditerranée et de la mer Rouge (Tunicata, Thaliacea). Beaufortia **22**(291), 83–103 (1974)

Gohar, H. A. F.: The place of the Red Sea between the Indian Ocean and the Mediterranean. Hidrobiol., Istanbul (Ser. B) **2**(2–3), 47–82 (1954)

Goldberg, M.: The geology of Tiran Island. Israel J. Earth Sci. **12**(2), 83–84 (1963)

Gorgy, S.: Les pêcheries et le milieu marin dans le secteur méditerranéen de la Republique Arabe Unie. Rev. Trav. Inst. (sci. tech.) Pêches Marit., **30**(1), 25–80 (1966)

Gottlieb, E.: A study of the benthos in Haifa Bay. Ecology and zoogeography of invertebrates. Ph. D. thesis, Hebrew Univ. Jerusalem, 1959 (in Hebrew)

Grace, S. F.: The influence of friction on the tidal motion of the Gulf of Suez. Roy. Astronom. Soc. Geophys. Suppl. II(7), 309–318 (1931)

*Gravier, Ch. J.: Sur les Annélides polychètes de la Mer Rouge. Bull. Mus. Hist. Natur., Paris (Sér. 1) **12**(1), 33–43; (2), 110–115 (1906a)

*Gravier, Ch. J.: Sur la faune annélidienne de la Mer Rouge et ses affinités. C. R. Acad. Sci. Paris, **142**(7), 410–412 (1906b)

Gravier, Ch. J.: Contribution a l'étude des Annélides polychètes de la Mer Rouge (Suite). Arch. Mus. Hist. Natur. Paris, (Sér. 4) **4**, (10), 67–168 (1908)

Gruvel, A.: Répartition géographique de quelques crustacés comestibles sur les côtes d'Egypte et de Syrie. Compt. Rend. Soc. Biogéogr., Paris **39**, 45–46 (1928)

Gruvel, A.: De l'influence du percement du Canal de Suez sur la faune marine des côtes de Syrie. Comm. Acad. Sci. Colon. Paris, **188**, 1697–1699 (24. 6. 1929)

Gruvel, A.: Les États de Syrie. Richesses marines et fluviales. Exploitation actuelle, avenir. Paris. Soc. Ed. Geogr. Marit. Colon. Franc. 1931, 453 pp.

Gruvel, A.: Sur quelques observations au sujet du Grand lac Amer (Canal de Suez). Comm. Acad. Sci. Paris (27. 6. 1932)

Gruvel, A.: Abondance du *Branchiostoma lanceolatum* Pallas dans le Canal de Suez. Comm. Acad. Sci. Paris, **197**, 92–93 (3. 7. 1933a)

Gruvel, A.: Sur la distribution de quelques éspèces de mollusques dans les lagunes du lac Timsah (Canal de Suez). Comm. Acad. Sci. Paris (10. 7. 1933b)

Gruvel, A.: Sur quelquesunes des causes qui arrêtent la pénétration des éspèces animales dans le Canal de Suez. Comm. Acad. Sci. Paris (26. 3. 1934)

Gruvel, A.: Contribution à l'étude de la Bionomie générale et de l'exploitation de la faune du Canal de Suez. Mém. Inst. Egypte, **29**, 1–255 (1936)

Gruvel, A., Chabanaud, P.: Poissons. Mission A. Gruvel dans le Canal de Suez. II. Mém. Inst. Egypte, **35**, 1–30 (1937)

*Gruvel, A., Moazzo, G.: Contribution à la Faune Malacologique marine des côtes Libano-Syriennes. In: Les états de Syrie. Gruvel, A. (ed.). Soc. Ed. Geogr. Maritim. Coloniales, Paris, 437–453 (1931)

Gruvel, A., Moazzo, G.: Mollusques téstacés du Grand lac Amer (Canal de Suez), (Première Liste). Bull. Mus. Hist. Natur. Paris, (Sér. 2) **5**, 143–147 (1933)

*Guérin-Meneville, F. E.: Crustacés. In: Brulle, A. Des animaux articulés. Expedition scientifique de Moree. Section des Sciences Physiques, **3**(1), 30–50 (1832)

Guille, A.: Sur la présence d'*Ophiactis savignyi* Müller et Troschel dans la région de Banyuls sur Mer. Vie et Milieu, A. **19**(2), 497–500 (1968)

Gurney, R.: XV. Report on the larvae of the Crustacea Decapoda. Cambridge Expedition to the Suez Canal, 1924. Trans. Zool. Soc. London **22**, 231–286 (1927a)

Gurney, R.: XXVI. Report on the Species of *Hippolyte*. Cambridge Expedition to the Suez Canal, 1924. Trans. Zool. Soc. London **22**, 391–397 (1927b)

Gurney, R.: VIII. Report on the Crustacea: Copepoda and Cladocera of the Plankton. Cambridge Expedition to the Suez Canal, 1924. Trans. Zool. Soc. London **22**, 139–172 (1927c)

Gurney, R.: IX. Report on the Crustacea: Copepoda of brine-pools at Kabret. Cambridge Expedition to the Suez Canal, 1924. Trans. Zool. Soc. London **22**, 173–177 (1927d)

Gurney, R.: XXXIII. Report on the Crustacea: Copepoda (Littoral and semi-parasitic). Cambridge Expedition to the Suez Canal, 1924. Trans. Zool. Soc. London **22**, 451–577 (1927e)

Haas, G.: Mollusca Marina. In: Bodenheimer, F. S. Prodromus Faunae Palaestinae. Mém. Inst. Egypte, 275–280 (1937)

Haas, G.: A Ctenophore from the Palestinian coast. Nature (London) **149**, 110–111 (1942)

Haas, G.: Sur l'immigration de Mollusques Indo-Pacifiques dans les eaux côtières de la Palestine. J. Conchyl. Paris **88**, 141–144 (1948)

Haas, G.: Preliminary report on the Molluscs of the Palestine coastal shelf. Bull. Sea Fish. Res. Stn. Haifa (1), 1–20 (1951)

Haas, G., Steinitz, H.: Erythrean fishes on the Mediterranean coast of Palestine. Nature (London) **160**(4053), 28 (1947)

Halim, Y.: Observations on the Nile bloom of phytoplankton in the Mediterranean. J. Cons. Perm. Int. Explor. Mer. **26**, 57–67 (1960)

Halim, Y.: Microplancton des eaux egyptiennes. Le genre *Ceratium* Schrank (Dinoflagellés). Rapp. Comm. Int. Explor. Mer Médit. **17**, 495–502 (1963)

Halim, Y.: Plankton of the Red Sea. Oceanogr. Mar. Biol. Ann. Rev. Barnes, H. (ed.). **7**, 231–275 (1969)

Halim, Y., Guergues, S, K., Saleh, H. H.: Hydrographic conditions and plankton in the south-east Mediterranean during the last normal Nile Flood (1964). Intern. Revue Ges. Hydrobiol. **52**(3), 401–425 (1967)

Hamel, H.: Les Caulerpes Méditerranéenes. Rev. Algol. **5**, 229–230 (1930)

Hamel, H.: Chlorophycées des côtes françaises (Ordre des Siphonales). Rev. Algol. **5**, 384–390 (1931)

Hammerton, E.: The Nile River—a case history. In: River Ecology and Man. London-New York: Academic Press, 1972, pp. 171–215

Harant, H.: XXIV. Rapport sur les Tuniciers. Cambridge Expedition to the Suez Canal, 1924. Trans. Zool. Soc. London **22**, 365–374 (1927)

Harlock, R., Laubier, L.: Notes on *Branchiosyllis uncinigera* (Hartmann-Schröder, 1960) new to the Mediterranean. Israel J. Zool. **15**, 18–25 (1966)

Harmelin, J. G.: Contribution à l'étude de l'Endofaune des Prairies d'*Halophila stipulacea* de Méditerranée Orientale. I. Annélides Polychètes. Rec. Trav. Stn. Mar. Endoume Fac. Sci. Bull. **45**(61), 305–316 (1969)

Harmer, S. F.: The Polyzoa of the Siboga Expedition. Pt. 4. Cheilostoma Ascophora, II. Rep. Siboga Exped. **28d**, 641–1147 (1957)

Hartmann-Schröder, G.: Polychaeten aus dem Roten Meer. Ergebnisse Nr. 8 der Reise von A. Remane und E. Schulz nach dem Roten Meer. Kiel. Meeresforsch. **16**(1), 69–125 (1960)

*Hartmeyer, R.: Über einige Ascidien aus dem Golf von Suez. S. B. Ges. Naturf. Freund. Berlin, 397–430 (1915)

Hastings, A. B.: XX. Report on the Polyzoa. Cambridge Expedition to the Suez Canal, 1924. Trans. Zool. Soc. London **22**, 331–354 (1927)

Hedgpeth, J. W.: Estuaries and Lagoons II. Biological aspects. In: Treatise on Marine Ecology and Palaeoecology. Hedgpeth, J. W. (ed.). Geol. Soc. Am. Memoir **67**, 49–54 (1957)

*Heller, C.: Die Crustaceen des südlichen Europa. Crustacea, Podophthalma. Mit einer Übersicht über die horizontale Verbreitung sämtlicher europäischer Arten, 1863, 338 pp. + 10 pl.

Heron-Allen, E., Earland, A.: II. Report on the Foraminifera. Cambridge Expedition to the Suez Canal, 1924. Trans. Zool. Soc. London **22**, 65–70 (1926)

Hérubel, M. A.: Sur les Sipunculides nouveaux rapportés de la Mer Rouge par M. Ch. Gravier (Note préliminère). Bull. Mus. Nat. Hist. Natur. **10**, 476–480 (1904)

Hildebrand, S. F.: The Panama Canal as a passageway for fishes, with lists and remarks on the fishes and invertebrates observed. Zoologica **24**(3), 15 pp. (1939)

Holthuis, L. B.: Notes on a collection of Crustacea Decapoda from the Great Bitter Lake, Egypt, with a list of the species of Decapoda known from the Suez Canal. Zool. Meded. Rijksmus. Nat. Hist. Leiden **34**(22), 301–330 (1956)

Holthuis, L. B., Gottlieb, E.: Two interesting crabs (Crustacea, Decapoda, Brachyura) from Mersin Bay, S. E. Turkey. Zool. Meded. Rijksmus. Nat. Hist. Leiden **34**(21), 287–298 (1956)

Holthuis, L. B., Gottlieb, E.: An annotated list of the Decapod Crustacea of the Mediterranean coast of Israel, with an appendix listing the Decapoda of the Eastern Mediterranean. Bull. Res. Coun. Israel, **7B**(1–2), 1–126 (1958)

*Hornell, J.: Report on the Fisheries of Palestine. Crown Agents for the Colonies. London: His Majesty's Stationary Office, 1935, 106 pp.

Horowitz, A.: The geology of Museri Island (Dahlak Archipelago, Southern Red Sea). Israel J. Earth Sci., **16**(2), 74–83 (1967)

Hsu, K.J., Ryan, W.B.F.: Report on Deep Sea Drilling in the Mediterranean. Rapp. Comm. Intern. Explor. Mer. Médit. **22**(2a), 19–23 (1973)

Hubbs, C.L.: The Suez Canal as a means in the dispersal of marine fishes. Copeia 162, 94 (1927)

Hughes, C., Kay, A.J.: Organisms as producers of carbonate sediment and indicators of environment in the Southern Persian Gulf. In: The Persian Gulf. Purser, B.H. (ed.) Berlin-Heidelberg-New York: Springer, 1973, pp. 33–56

Hume, W.F.: The topography and geology of the Peninsula of Sinai (southeastern portion). Egypt. Survey Dept., Cairo, 1–280 (1906)

*Hume, W.F.: Report on the oilfields region of Egypt. Egypt. Surv. Dept. Cairo: Govt. Press. 1916, I-VIII, 1–103

Huvé, H.: Sur une variété, nouvelle pour la méditerranée, du *Caulerpa racemosa* (Forsskal) Agardh. Rec. Trav. Stn. Mar. Endoume. Fasc. 21 (Bull. 12), 67–73 (1957)

Ingle, R.W.: Crustacea Stomatopoda from the Red Sea and Gulf of Aden. Bull. Sea Fish. Res. Stn., Haifa, (33), 1–69 (1963)

Issel, A.: Della Fauna malacologica del Mar Rosso. Bol. Soc. Geogr. Italiana Pisa, 387 pp. (1869)

*Issel, A.: Essai sur l'origine et la formation de la Mer Rouge. Bull. Soc. Belge Géol. Paléont. & Hydrol. (sér. 2), **13**, 65–84 (1900)

Jackson, C.H.N.: XVII. Report on a new species of Collembolan. Cambridge Expedition to the Suez Canal, 1924. Trans. Zool. Soc. London **22**, 289–290 (1927)

James, D.B., Pearse, J.S.: Echinoderms from the Gulf of Suez and the northern Red Sea. J. Mar. Biol. Assoc. India **11**, 78–125 (1969)

Jordan, D.S., Hubbs, C.L.: Notes on a collection of fishes from Port Said, Egypt. Ann. Carneg. Mus. **11** (18), 461–468 (1917)

Kalugina-Gutnik, A.A.: Species composition and geographical distribution of the Red Sea Macrophyta. In: The Benthos of the Red Sea Shelf. Vodyanitskii, V.A. (ed.). Akademia Nauk U.S.S.R. Kiev: Naukova Dumka, 1971, pp. 232–268 (in Russian)

Keller, C.: Die Fauna im Suez-Kanal und die Diffusion der Mediterranen und Erythräischen Tierwelt. Neue Denkschr. Schweiz. Ges. f. Naturforsch., Zürich (Ser. 3) **28**, 1–39 (1882)

Keller, C.: Naturwissenschaftliche Reisen auf dem Isthmus und am Roten Meere. Mitt. Ostschw. Geogr. Gesell. **1**, 1–30 (1883)

Keller, C.: Die Wanderung der marinen Tierwelt im Suezkanal. Zool. Anzeig., Leipzig, **11**(283), 359–364; (284), 380–395 (1888)

*Keller, C.: Das Leben des Meeres. Leipzig: Wilhelm Engelman 1895, pp. 159–171

Kemp, S.: Pontoniinae. Notes on Crustacea Decapoda in the Indian Museum. XVII. Rec. Indian Museum, Calcutta, **27**, 249–343 (1922)

Kessler, P.: The structural and geomorphic evolution of the Persian Gulf. In: The Persian Gulf. Purser, B.H. (ed.). Berlin-Heidelberg-New York: Springer, 1973

Khmeleva, N.N.: The role of radiolarians in the estimation of the primary production in the Red Sea and the Gulf of Aden. Doklady Akad. Nauk SSSR, **172**(6), 1430–1433 (1967) (in Russian)

Kimor, B.: The Suez Canal as a link and a barrier in the migration of planktonic organisms. Israel J. Zool. **21**(3–4), 391–403 (1972)

Kimor, B.: Plankton relations of the Red Sea, Persian Gulf and Arabian Sea. In: The Biology of the Indian Ocean, Ecological Studies Zeitschel, B. (ed.). Berlin-Heidelberg-New York: Springer, 1973, Vol. III, pp. 221–232

Kimor, B., Berdugo, V.: Preliminary report on the plankton of Bardawil—a superhaline lagoon in northern Sinai. In: The Hebrew University—Smithsonian Institution Joint Program: Biota of the Red Sea and the Eastern Mediterranean. Interim Report. Dept. Zool., Hebrew Univ. Jerusalem, 1969, pp. 90–96 (mimeo)

Kingdon, J.: East African Mammals—An Atlas of Evolution in Africa. London-New York: Academic Press, 1971, 446 pp.

Kinsman, J.J.: Reef coral tolerance of high temperatures and salinities. Nature (London) **202**, 1280–1282 (1964)

Kiseleva, M.I.: Qualitative composition and quantitative distribution of the Polychaeta of the Red Sea. In: The Benthos of the Red Sea Shelf. Vodyanitskii, V.A. (ed.). Akademia Nauk U.S.S.R. Kiev: Naukova Dumka, 1971, pp. 44–76 (in Russian)

Kiseleva, M. I., Murina, V. V.: Conclusions. In: The Benthos of the Red Sea Shelf. Vodyanitskii, V. A. (ed.). Akademia Nauk U.S.S.R. Kiev: Naukova Dumka, 1971, pp. 268—272 (in Russian)

Kiseleva, M. I., Tchukhtchin, V. D.: Some data on the quantitative development of the macro- and meiobenthos in the Eastern Mediterranean. Akademia Nauk S.S.S.R. Okean, K., 1965, 192 pp. (in Russian)

Klausewitz, W.: Systematisch-evolutive Untersuchungen über die Abstammung einiger Fische des Roten Meeres. Z. Anz. Supplementband 23, 175–182 (1960)

Klausewitz, W.: Die Erforschung der Ichthyofauna des Roten Meeres. Foreword to: Klunzinger, C. B. Synopsis der Fische des Roten Meeres. Weinheim: J. Cramer, 1964, pp. 3–35

Klausewitz, W.: Remarks on the zoogeographical situation of the Mediterranean and the Red Sea. Annal. Mus. Civ. Stor. Nat. Giacomo Doria, **77**, 323–328 (1968)

Klausewitz, W.: New aspects of the palaeogeography of the Mediterranean ichthyofauna. Ichthyologia (Belgrad) **5**(1), 79–86 (1973)

* Klunzinger, C. B.: Die Fische des Rothen Meeres. Stuttgart: E. Schweizerbart, 1884

* Kobelt, W.: Keller's Beobachtungen am Suez Canal. Nachrbl. deut. Malak. Ges. **15**, 117 (1883)

Komarovsky, B.: A cladoceran from the plankton as a possible indicator for the presence of the Nile flood off the Israeli coast. Nature (London) **171** (937), 464–465 (1953)

Komarovsky, B.: The Tintinnina of the Gulf of Eylath (Aqaba). Bull. Sea Fish. Res. Stn. Haifa, (21), 1–40 (1959)

Kossmann, R.: Zoologische Ergebnisse einer im Auftrage der Königlichen Akademie der Wissenschaften zu Berlin ausgeführten Reise in die Küstengebiete des Rothen Meeres. Leipzig: Wilhelm Engelmann, 1877, pp. 1–273

Kosswig, C.: Die Faunengeschichte des Mittel- und Schwarzen Meeres. Compt. Rend. Soc. Turque Sci. Phys. Natur. **9**, 37 (1942)

Kosswig, C.: Über Tethysrelikte in der türkischen Fauna. Compt. Rend. Soc. Turque Sci. Phys. Natur. **10**, 31 (1943)

Kosswig, C.: Erythräische Fische im Mittelmeer und an der Grenze der Ägais. Syllegomena Biologica. Festschrift Kleinschmidt. Leipzig: Akademie Verlag 1950, pp. 203–212

Kosswig, C.: Beitrag zur Faunengeschichte des Mittelmeeres. Publ. Staz. Zool. Napoli, **28**, 78–88 (1956)

Kosswig, C.: Tethys and its relation to the Peri-Mediterranean Faunas of freshwater fishes. In: Aspects of Tethyan Biogeography Adams, C. G., Ager, D. V. (eds.). Syst. Assoc. Publ. No. **7**, 313–324 (1967)

Krauss, W.: Eine Bemerkung zur Abnahme des Salzgehaltes im Suezkanal. Deut. Hydrogr. Z. **11**(3), 109–112 (1958)

Krefft, G.: *Platycephalus indicus* (L.), 1758, ein neues Faunenelement der ägyptischen Mittelmeerküste. Arch. Fisch. Wiss. **14**(3), 148–152 (1963)

Krukenberg, C. Fr. W.: Die Durchflutung des Isthmus von Suez in chorologischer, hydrographischer und historischer Beziehung. Heidelberg: Carl Winter's Univ. Buchhandlung, 1888a, 156 pp.

Krukenberg, C. Fr. W.: Wissenschaftliche Ergebnisse einer Reise vom Etang de Berre über Marseille und Triest nach Suakim und Massaua. Vergl. Physiol. Studien. 2 Reihe, 5 Abt., II Teil, Heidelberg: 1888b, pp. 9–27

Ktari-Chakroun, F., Bahloul, M.: Capture de *Siganus luridus* (Rüppell) dans le golfe de Tunis. Bull. Inst. Nat. Sci. Technol. Oceanogr. Pêche Salambô **2**(1), 49–52 (1971)

Lacombe, H., Tchernia, P.: Quelques traits generaux de hydrologie Méditerranéenne d'après divers campagnes hydrologiques récentes en Méditerranée dans le prôche Atlantique et dans le detroit de Gibraltar. Cah. Oceanogr. **12**(8), 527–548 (1960)

Lami, R.: Quelques algues du grand lac amer (Basse-Egypte) récoltées par M. le Professeur Gruvel, en Avril 1932. Rev. Algol., **6**, 355–356 (1932)

Larwood, H. J.: The Fishery Grounds near Alexandria XXI. Tanaidacea and Isopoda. Notes Fish. Res. Dir. Cairo (35), 1–72 (1940)

Larwood, H. J. C.: LXIX. Crustacea Tanaidacea and Isopoda from the Suez Canal. Ann. Mag. Nat. Hist. London (Ser. 12) **7**(80), 561–577 (1954)

Laskaridis, K.: A revision of the goatfish genus *Upeneus* with descriptions of two new species. Proc. U.S. Nat. Mus. **103**(3330), 497–532 (1948a)

Laskaridis, K.: *Holocentrum rubrum* (Forsk.) and *Lagocephalus lagocephalus* (L.), two newly reported members of the Greek fish-fauna (Dodecanesian Islands). Praktika Hellen. Hydrobiol. Inst. **2**(1), 127–129 (1948b)

Lattin, G. de: Grundriß der Zoogeographie. Stuttgart: Gustav Fischer, 1967, 602 pp.

Laubier, L.: Sur quelques Annélides polychètes de la région de Beyrouth. Misc. Pap. Nat. Sci. Am. Univ. Beirut, **5**, 9–22 (1966)

Laurie, R. D.: On the Brachyura. Reports on the marine biology of the Sudanese Red Sea. XXI. J. Linn. Soc. London (Zool.) **31**, 407–475 (1915)

Lawrence, J. M.: Temperature tolerances of tropical shallow-water echinoids (Echinodermata) at Elat (Red Sea). Israel J. Zool. **22**, 143–150 (1973)

Le Danois, E.: Recherches sur les fonds chalutables des côtes de Tunisie. Ann. Stat. Ocean. Salammbô 1. 1925

Leigh-Sharpe, W. H.: X. Report on a parasitic copepod of *Atherina pinguis* (*Lernaeanicus cerberus* sp.n.) Cambridge Expedition to the Suez Canal, 1924. Trans. Zool. Soc. London **22**, 179–183 (1927)

Leloup, E.: Amphineures du Golfe d'Aqaba et de la peninsule Sinai. Bull. Sea Fish. Res. Stn., Haifa, (29), 25–55 (1960)

Lerner-Seggev, R.: Ostracoda of the Suez Canal. Israel J. Zool. **21** (3–4), 239–248 (1972)

Lesseps, F. de: Communication sur les lacs amers de l'Isthme de Suez. I. Comm. Acad. Sci. Paris **78**, 1740–1748 (1871)

* Lesseps, F. de: Reponse de M. de Lesseps aux observations de M. de Lanoue sur les ensablements a Port Said et sur la salure des Lacs Amers. Bull. Inst. Egypte, **12**, 20–23 (1873)

Lesseps, F. de: Deuxième communication sur les lacs amers de l'Isthme de Suez. Comm. Acad. Sci. Paris **82**, 1133–1138 (1876a)

Lesseps, F. de: Etude de plusieurs questions relatives au canal de Suez. Compt. Rend. Acad. Sci. Paris, 1137 (1876b)

Leston, D.: Spread potential and colonisation of islands. Syst. Zool. **6**, 1, 41–46 (1957)

Letronne, A. J.: L'Isthme de Suez: le canal de jonction des deux mers. Revue des Deux-Mondes (15 juillet), 1841

Levi, C.: Spongiaires des côtes d'Israël. Bull. Res. Coun. Israel **6B**(3–4), 201–212 (1957)

Levi, C.: Spongiaires de la Mer Rouge recueillis par la "Calypso" (1951–52). Annal. Inst. Océanogr., Monaco, **34**, 3–46 (1958)

Levi, C.: Spongiaires récoltés par l'Expedition Israelienne dans le Sud de la Mer Rouge en 1962. Bull. Sea Fish. Res. Stn., Haifa, (40), 3–27 (1965)

Lewinsohn, Ch., Holthuis, L. B.: New records of Decapoda Crustacea from the Mediterranean coast of Israel and the Eastern Mediterranean. Zool. Meded. Rijks. Mus. Nat. Hist. Leiden, **40**(8), 45–63 (1964)

Liebmann, E.: Contribution to the knowledge of Palestine sea fishes. Rapp. P.-v. Comm Intern. Explor. Mer Médit. **8**, 317–327 (1934)

Liebmann, E.: Oceanographic observations on the Palestine coast. Rapp. P.-v. Comm. Int. Explor. Mer Médit. **9**, 181–185 (1935)

Lindroth, C. H.: Ecological zoogeography. Ann. State Res. Coun. Nat. Sci. Stockholm 1955, 199–203 (1956)

Lindroth, C. H.: The faunal connections between Europe and America. Stockholm: Almquist, Wiksell and Willey, 1957, 344 pp.

Lipkin, Y.: Marine algal and sea-grass flora of the Suez Canal (The significance of this flora to the understanding of the recent migration through the Canal). Israel J. Zool. **21**(3–4), 405–446 (1972a)

Lipkin, Y.: Vegetation of the Bitter Lakes in the Suez Canal water system. Israel J. Zool. **21**(3–4), 447–457 (1972b)

Lipkin, Y.: *Halophila stipulacea*, a review of a successful immigration. Aquatic Bot. **1**, 203–215 (1975)

Lipkin, Y., Safriel, U.: Intertidal zonation on rocky shores at Mikhmoret (Mediterranean, Israel). J. Ecol., **59**, 1–30 (1971)

* Lisitzin, E.: The mean sea level of the world ocean. Commentat. Phys.-Math. Helsingf. **30**(7), 1–35 (1965)

Lisitzin, E.: Sea-Level Changes. Elsevier Oceanography Ser. 8, Amsterdam: Elsevier, 1974, 286 pp.

Lissner, H.: Sardine fishing in Israel. Sci. Techn. Informat. Sea Fish. Res. Stn., Haifa (2), 1–25 (1949) (mimeo)

Logan, B.W.: *Cryptozoon* and associate stromatolites from the recent Sharks Bay Western Australia. J. Geol. **69**(5), 517–533 (1961)

Loya, Y.: The Red Sea coral *Stylophora pistillata* is an *r* strategist. Nature (London) **259**(5542), 478–480 (1976)

Lourie, A., Ben-Tuvia, A.: Two Red Sea fishes *Pelates quadrilineatus* (Bloch) and *Crenidens crenidens* (Forsskal) in the Eastern Mediterranean. Israel J. Zool. **19**(4), 203–207 (1970)

Luksch, J.: Expedition S.M. Schiff „Pola" in das Rothe Meer, nördliche Hälfte (Oct. 1895–Mai 1896). Wissenschaftliche Ergebnisse. VI. Physikalische Untersuchungen. Denkschr. Akad. Wiss., (Math. Naturw. Kl.) Wien **65**, 351–422 (1898)

Lyle, L.: Algae of the Suez Canal. J. Botany (London) **68**, 327 (1930)

Mac Andrew, R.: Report on the testaceous Mollusca obtained during a dredging excursion in the Gulf of Suez in the months of February and March 1869. Annal. Mag. Natur. Hist., London (Ser. 4) **6**(36), 429–450 (1870)

McCosker, J.D., Dawson, C.E.: Biotic Passage through the Panama Canal, with Particular Reference to Fishes. Mar. Biol. **30**, 343–351 (1975)

MacDonald, R.: An examination of plankton hauls made in the Suez Canal during the year 1928. Notes Mem. Fish. Res. Dir. Cairo, (3), 1–11 (1933)

McIntyre, A.: The Coccolithophoridae in Red Sea sediments. In: Hot brines and recent heavy metal deposits in the Red Sea. Degens, E.T., Ross, D.A. (eds.) Berlin-Heidelberg-New York: Springer, 1969, pp. 299–305

Makkaveeva, E.B.: Qualitative composition and quantitative distribution of Red Sea Tanaidacea. In: The Benthos of the Red Sea Shelf. Vodyanitskii, V.A. (ed.). Akademia Nauk U.S.S.R. Kiev: Naukova Dumka, 1971, pp. 88–109 (in Russian)

Mars, P.: Les faunes et la stratigraphie du Quaternaire méditerranéen. Rec. Trav. Stn. Mar. Endoume, **28**(43), 61–97 (1963)

Mars, P., Picard, J.: Note sur les gisements sous-marins a faune céltique en Méditerranée. Rapp. Comm. Int. Explor. Mer Médit., **15**, 325–336 (1958)

* Martens, E. von: Conchylien aus dem Suez Canal. Sitzungsber. Ges. Naturforsch. Freunde, Berlin (6), 94 (1887)

Maurin, C.: Ecologie ichthyologique des fonds chalutables Atlantiques (de la baie iberomarocaine a la Mauritanie), et de la Méditerranée occidentale. Thesis, Fac. Sci. Univ. Nancy, 1968, 146 pp.

Mayr, E.: What is a Fauna? Zool. Jb. Syst. **92**, 473–486 (1965a)

Mayr, E.: Evolution at the species level. In: Ideas in Modern Biology. Moore, J.A. (ed.). New York: Natural History Press, 1965b, pp. 315–325

Mendelssohn, H.: A new locality for *Cyprinodon dispar* Rüppell. Nature (London) **160**, 123 (1947)

Menzies, J.R.: Transport of marine life between oceans through the Panama Canal. Nature (London) **220**(5169), 802–803 (1968)

Mienis, H.K.: On the nomenclature and distribution of *Columbella savignyi* (Gastropoda, Columbellidae). Argamon, J. Israel Malac. Soc. **3**, 17–19 (1972)

Mienis, H.K.: Notes of Recent and Fossil Neritidae. 5. *Neritina kinzelbachi* Nordsieck, a Synonym of *Nerita sanguinolenta* Menke. Argamon J. Israel Malac. Soc. **4**, 45–46 (1974)

Miller, A.R., Munns, R.G.: The Bitter Lake salt barrier. In: l'Océanographie Physique de la Mer Rouge. Symp. Assoc. Int. Sci. Phys. Océan. CNEXO, Serie: Actes Coll. **2**, 295–309 (1974)

* Mitchell, J.C.: Rapport sur les poissons commestibles du lac Menzaleh et les procédes en usage pour les pêcher et les conserver. Cairo: 1895

Moazzo, P.G.: Mollusques testacés marins du Canal de Suez. Mém. Inst. Egypte **38**, 1–283 (1939)

Monod, Th.: Une expedition scientifique au Canal de Suez. Rev. Gen. Sci. **37**(13), 387 (1925)

Monod, Th.: Über einige Indo-Pazifische Decapoden der Meeresfauna Syriens. Zool. Anzeig. Leipzig **92**, (5–6), 135–141 (1930)

Monod, Th.: Crustacées exotiques en Méditerranée. La Terre et la Vie (Rev. Hist. Nat.) Paris: 1932, Vol. II, pp. 65–73

Monod,Th.: Tanaidacea et Isopoda. In: Mission Robert Ph. Dollfus en Egypte (Decembre 1927–Mars 1929). Mém. Inst. Egypte **21**, 161–264 (1933)

Monod,Th.: Crustacés. Mission A. Gruvel dans le Canal de Suez. I. Mém. Inst. Égypte, **34**, 1–19 (1937)

Monod,Th.: Decapoda Brachyura. In: Mission Robert Ph. Dollfus en Égypte. VIII. Mém. Inst. Egypte **37**, 91–162 (1938)

Monro,C.C.A.: A note on a collection of Polychaeta from the Eastern Mediterranean, with a description of a new species. Ann. Mag. Natur. Hist., London (Ser. 10) 19, 82–86 (1937)

Monterosato,T., Marquis de: Coquilles marines de Chypre. J. Conchyliol. **67**(4), 1–10 (1899)

Morcos,S.A.: Die Verteilung des Salzgehaltes im Suez-Kanal. Kiel. Meeresforsch. **16**(2), 133–154 (1960)

Morcos,S.A.: Effect of the Aswan High Dam on the current regime in the Suez Canal. Nature (London) **214**(5091), 901–902 (1967a)

Morcos,S.A.: The chemical composition of sea water from the Suez Canal Region. Part I: The major anions. Kiel. Meeresforsch. **23**, 80–90 (1967b)

Morcos,S.A.: The chemical composition of sea water from the Suez Canal Region. Part II: Major cations. Kiel. Meeresforsch. **24**, 66–84 (1968)

Morcos,S.A.: Physical and Chemical Oceanography of the Red Sea. In: Oceanogr. Mar. Biol. Ann. Rev., Barnes,H. (ed.). **8**, 73–202 (1970)

Morcos,S.A.: Early Investigations of the Suez Canal Waters during and after its Opening in 1869. Proc. Roy. Soc. Edinburgh (B)**72** (45), 449–458 (1971/72).

Morcos,S.A.: A transitional stage in the current regime in the Suez Canal. Limnol. Oceanogr. **20**(4), 672–679 (1975)

Morcos,S.A., Gerges,M.A.: Circulation and mean sea level in the Suez Canal. In: L'Océanographie physique de la Mer Rouge, IAPSO-UNESCO-SCOR Symp., Paris, 1972. CNEXO Publ. Ser. Actes Colloq. 2. pp. 267–287 (1974)

Morcos,S.A., Messieh,S.N.: Circulation and salinity distribution in the southern part of the Suez Canal. Limnol. Oceanogr. **18**(1), 121–130 (1973a)

Morcos,S.A., Messieh,S.N.: Change in the Current regime in the Suez Canal after construction of Aswan High Dam. Nature (London) **242**(5392), 38–39 (1973b)

Morcos,S.A., Riley,J.P.: Chlorinity, salinity, density and conductivity of sea water from the Suez Canal region. Deep Sea Res., **13**(4), 471–479 (1966)

Mortensen,Th.: VI. Report on the Echinoderms. Cambridge Expedition to the Suez Canal, 1924. Trans. Zool. Soc. London **22**, 117–131 (1926)

Mortensen,Th.: Echinoderms from the Iranian Gulf. Asteroidea, Ophiuroidea and Echinoidea (with section on Crinoidea by Torsten Gislen). Danish Sciet. Invest. Iran, Pt. 2, 55–112 (1940)

Mortensen,Th., Steuer,A.: The Fishery Grounds near Alexandria. XIII. Echinoderms. Notes Hydrobiol. Fish. Res. Dir. Cairo (21), 1–37 (1937)

Moshkovitz,S.: The Mollusca in the upper part of the "Sakiebeds" (Upper Neogene—Lower Pleistocene) in the Central Coastal Plain of Israel. Israel J. Earth Sci. **12**(3), 97–146 (1963)

Moshkovitz,S.: The Mollusca in the marine Pliocene and Pleistocene sediments of the south-eastern Mediterranean basin (Cyprus-Israel). Ph. D. thesis, Hebrew Univ. Jerusalem. 1968, 153 pp., (in Hebrew with English summary)

* Müller,K.: Das Wasser des Bittersees im Suez-Kanal. Petermanns Mitt., 1872, Vol. XVIII, p. 191

Murina,V.V.: The qualitative und quantitative characteristics of the Red Sea macrobenthos. In: The Benthos of the Red Sea Shelf. Vodyanitskii,V.A. (ed.). Akademia Nauk U.S.S.R. Kiev: Naukova Dumka, 1971, pp. 3–22 (in Russian)

Muromtsev,A.M.: Hydrological conditions of the Suez Canal. Doklady Akad. Nauk U.S.S.R. **135**(1), 179–181 (1960) (in Russian)

Muromtsev,A.M.: Contributions to the hydrology of Suez Canal, Red Sea and Gulf of Aden. Meteorol. Hydrolog. Moscow, **2**, 42–45 (1962) (in Russian)

Muschler,R.: Enumération des algues marines et d'eau douce observées jusqu'à ce jour en Égypte. Mém. Inst. Egypte **5**, 141–237 (1908)

Nasr,A.H.: Synopsis of the marine algae of the Egyptian Red Sea coast. Bull. Fac. Sci. Egypt. Univ. No. 26, 1–155 (1947)

*Natterer, K.: Expedition S. M. Schiff „Pola" in das Rothe Meer, nördliche Hälfte (October 1895–Mai 1896). IX. Chemische Untersuchungen. Denkschr. Akad. Wissensch. (Math. Naturw. Kl.) Wien, **65**, 445–572 (1898)

Nesteroff, W. D.: La crise de salinité messinienne en Méditerranée; enseignements des forages JOIDES et du bassin de Sicilie. Rapp. Comm. Int. Explor. Mer Médit. **22**(2a), 29–31 (1973a)

Nesteroff, W. D.: Mineralogy, petrography, distribution and origin of the Messinian Mediterranean evaporites. Initial Reports DSDP **13**(2), 673–694 (1973b)

Neumann, A. C.: Red Sea. In: The Encyclopedia of Oceanography. Fairbridge, R. W. (ed.). New York: Van Nostrand Reinhold, 1966, pp. 748–751

Newman, W. A.: The National Academy of Science Committee on the Ecology of the Interoceanic Canal. In: The Panamic Biota. Some Observations Prior to a Sea-Level Canal. Jones, M. L. (ed.). Bull. Biol. Soc. Wash. **2**, 247–261 (1972)

Nilsson-Cantell, G. A.: Notes on a *Balanus* from the saline lake Birket el Gessabaia. Proc. Egypt. Acad. Sci. **4**, 43–44 (1949)

Nobili, G.: Tre nuovi sferomidi eritrei del Museo Zoologico dell' Universita di Napoli. Annu. Mus. Zool. R. Univ. Napoli Nuov. Ser. **2**(16), 1–7 (1906)

Nordsieck, F.: Marine Gastropoden aus der Shiqmona-Bucht in Israel. Arch. Moll. **102**, 227–245 (1972)

Norman, J. R.: XXV. Report on the Fishes. Cambridge Expedition to the Suez Canal, 1924. Trans. Zool. Soc. London **22**, 375–390 (1927)

Norman, J. R.: Note on the fishes of the Suez Canal. Proc. Zool. Soc. London **1929**(4), 615–616 (1929)

O'Donoghue, C. H.: XXXVIII. Report on the Opisthobranchiata. Cambridge Expedition to the Suez Canal, 1924. Trans. Zool. Soc. London **22**, 713–841 (1929)

Omer-Cooper, J.: XII. Report on the Crustacea Tanaidacea and Isopoda. Cambridge Expedition to the Suez Canal, 1924. Trans. Zool. Soc. London **22**, 201–209 (1927)

Oren, O. H.: Some hydrographical features observed off the coast of Israel. Bull. Inst. Océanogr. Monaco (1017), 1–9 (1952)

Oren, O. H.: Changes in temperature of the Eastern Mediterranean Sea in relation to the catch of the Israel trawl fishery during the years 1954/55 and 1955/56. Bull. Inst. Océanogr. Monaco **1102**, 1–12 (1957)

Oren, O. H.: "Chypre-02" croisière dans la Méditerranée Orientale. Données Oceanographiques. Cah. Oceanogr. 18, Suppl. 1, 1–17 (1966)

Oren, O.H.: Oceanographic and biological influence of the Suez Canal, the Nile and the Aswan Dam on the Levant basin. Rep. Progr. Oceanogr. **5**, 161–167 (1969)

Oren, O. H.: The Suez Canal and the Aswan High Dam—their effect on the Mediterranean. Underwater Sci. Technol. J., Dec. 1970, **2**, 222–229 (1970)

Oren, O. H., Hornung, H.: Temperatures and salinities off the Israel Mediterranean Coast. Bull. Sea Fish. Res. Stn. Haifa (59), 17–31 (1972)

Oren, O. H., Komarovsky, B.: The influence of the Nile flood on the shore waters of Israel. Rapp. Comm. Int. Explor. Mer Médit. **16**(3), 655–659 (1961)

*Paget, G. W.: Report on the Fisheries of Egypt for the years 1920, 1921, 1922. Govt. Press, Cairo (1921–1923)

Paldi, R.: Computer produced regional bibliography (Eastern Mediterranean, Suez Canal and Red Sea). Bull. Sea Fish. Res. Stn., Haifa (57), 1–17 (1970)

Pallary, P.: Catalogue des Mollusques du littoral Méditerranéen de l'Égypte. Mém. Inst. Egypte, le Caire **7**(3), 69–205 (1912a)

Pallary, P.: Liste des mollusques marins des côtes de la Syrie. Feuille des Jeunes Naturalistes, Paris (sér. 5) **42** (504), 171–174 (1912b)

Pallary, P.: Enumération des mollusques marins des côtes de Syrie. Bull. Soc. Hist. Natur. Afrique Nord. Alger, juillet 165–172; nov. 175–178 (1919)

Pallary, P.: Les mollusques marins de la Syrie. J. Conchyl. Paris **82**(1), 5–58 (1938)

*Palombi, A.: Notizie faunistiche sul Canale di Suez. Ann. R. Liceo Scient. Avellino, p. 94 (1927)

Palombi, A.: Report on the Turbellaria. Cambridge Expedition to the Suez Canal, 1924. Trans. Zool. Soc. London **22**, 579–631 (1928)

Papenfuss, G. F.: A history, catalogue, and bibliography of Red Sea benthic algae. Israel J. Botan. **17**(1–2), 1–118 (1968)

Paperna, I.: Parasitic Helminths of Inland Water Fishes in Israel. Israel J. Zool. **13**, 1–20 (1964)

Paperna, I.: Monogenea from Red Sea fishes. I. Monogenea of fish of the genus *Siganus* Proc. Helminthol. Soc. Wash. **39**(1), 33–39 (1972a)

Paperna, I.: Monogenea from Red Sea fishes. II. Monogenea of Mullidae. Proc. Helminthol. Soc. Wash. **39**(1), 39–45 (1972b)

Paperna, I.: Monogenetic trematodes of cyprinodont fishes in the Near East. Zool. Anz. Leipzig **188**, 114–116 (1972c)

Paperna, I.: Parasitological implications of fish migration through interoceanic canals. 17e Congr. Intern. Zool. (Monte Carlo, Sept. 1972). Theme No. 3: Les conséquences biologiques des canaux inter-océans. 9 pp. (1972d) (mimeo)

Paperna, I.: Parasites and diseases of the Grey Mullet (Mugilidae) with special reference to the seas of the Near East. Aquaculture **5**, 65–80 (1975)

Paperna, I., Lahav, M.: Parasites of fish of the hypersaline Bardawil Lagoon, North Sinai—A preliminary communication. Rapp. Comm. Int. Explor. Mer Médit. **23**(3), 127–128 (1975)

Paperna, I., Overstreet, R. M.: Parasites and Diseases of Grey Mullets (Mugilidae). Cambridge: Univ. Press (IBP series), (in press) 118 pp.

Parenzan, P.: Carta d'identita delle conchiglie del Mediterraneo. Vol. I. Gasteropodi. Taranto: Ed. Bios-Taras, 1970, 243 pp.

Parker, F. L.: Eastern Mediterranean Foraminifera. Report on the Swedish Deep-Sea Exped. 1947–1948, Stockholm, **8**, 217–283 (1958)

Pearse, J. S.: Patterns of reproductive periodicities in four species of Indo-Pacific Echinoderms. Proc. Indian Acad. Sci. **68**, 247–279 (1968)

Pearse, J. S.: Reproductive periodicity of Indo-Pacific invertebrates in the Gulf of Suez. I. The echinoids *Prionocidaris baculosa* (Lamarck) and *Lovenia elongata* (Gray). Bull. Mar. Sci. **19**(2), 323–350 (1969)

* Pérès, J. M.: Contribution a l'étude des Ascidies de Tunisie. Bull. Stn. Océanogr. Salambô 49 (1954)

Pérès, J. M.: Ascidies recoltées sur les côtes Méditerranéennes d'Israel. Bull. Res. Coun. Israel **7B**(3–4), 143–150 (1958a)

Pérès, J. M.: Ascidies de la Baie de Haifa collectées par E. Gottlieb, Bull. Res. Coun. Israel **7B**(3–4), 151–164 (1958b)

Pérès, J. M.: The Mediterranean Benthos. Oceanogr. Mar. Biol. Ann. Rev. Barnes, H. (ed.). **5**, 449–533 (1967)

Pérès, J. M., Picard, J.: Nouveau Manuel de Bionomie Benthique de la Mer Méditerranée. Rec. Trav. Stat. Mar. Endoume **31**(47), 1–137 (1964)

Pesta, O.: Die Decapodenfauna der Adria. Versuch einer Monographie. Leipzig-Wien: 1918, 500 pp.

Petersson, H.: Reports of the Swedish Deep-Sea Expedition. Stockholm **1**, 1–123 (1957)

Pfannenstiel, M.: Quartäre Spiegelschwankungen des Mittelmeeres und des Schwarzen Meeres. Vierteljahrsschr. Naturforsch. Ges. Zürich **96**(2), 81–102 (1951)

Pfannenstiel, M.: Erläuterungen zu den bathymethrischen Karten des östlichen Mittelmeeres. Bull. Inst. Océanogr. Monaco 1192 (1960)

Pianka, E. R.: Latitudinal gradients in species diversity: a review of concepts. Am. Naturalist **100**, 33–46 (1966)

Picard, L.: Structure and evolution of Palestine, with comparative notes on neighbouring countries. Bull. Geol. Dept. Hebrew Univ. Jerusalem **2**(2–4), 187 pp. (1943)

* Picotti, M.: Ricerche di oceanographia chimica I. Tabelli generali delle analisi clorometriche e dei dati de temperatura, salinita e densita. Campagna Idrografica nel Mar Rosso della R.N. "Ammiraglio Magnaghi" 1923–1924. Annal. Idrografici 11, Genova, Inst. Idrog. Regia Marina (1927)

Por, F. D.: A study of the Levantine and Pontic Harpacticoida (Crustacea, Copepoda). Zool. Verh. Rijksmus. Nat. Hist. Leiden **64**, 1–128 (1964)

Por, F. D.: The Canuellidae (Copepoda, Harpacticoida) in the waters around the Sinai Peninsula and the problem of "Lessepsian" migration of this family. Israel J. Zool. **18**, 169–178 (1969a)

Por, F. D.: Limnology of the heliothermal Solar Lake on the coast of Sinai (Gulf of Elat). Verh. Int. Verein Theor. Angew. Limnol. **17**, 1031–1034 (1969b)

Por, F. D.: The zoobenthos of the Sirbonian lagoons. Rapp. Comm. Int. Explor. Mer Médit. **20**(3), 247–249 (1971a)

Por, F. D.: One hundred years of Suez Canal—A century of Lessepsian migration: retrospect and viewpoints. Syst. Zool. **20**, 138–159 (1971b)

Por, F. D.: Hydriobiological notes on the high-salinity waters of the Sinai Peninsula. Mar. Biol. **14**(2), 111–119 (1972)

Por, F. D.: The benthic Copepoda of the Sirbonian Lagoon (Sabkhat el Bardawil). Cah. Biol. Mar. **14**, 89–107 (1973a)

Por, F. D.: The nature of the Lessepsian migration through the Suez Canal. Rapp. Comm. Int. Explor. Mer Médit. **21**(9), 679–682 (1973b)

Por, F. D.: The stability of the Biota of the Bitter Lakes. Rapp. Comm. Int. Explor. Mer Médit. **22**(4), 111–112 (1973c)

Por, F. D.: Pleistocene pulsation and preadaptation of biotas in mediterranean seas: consequences for Lessepsian migration. Syst. Zool. **24**, 72–78 (1975a)

Por, F. D.: An outline of the Zoogeography of the Levant. Zoologica Scripta **4**, 5–20 (1975b)

Por, F. D.: Boring species of *Aspidosiphon* (Sipuncula) on the coasts of Israel. Proc. Intern. Symp. Biol. Sipuncula and Echiura I (Kotor 1970), Belgrade, 301–304 (1975c)

Por, F. D., Aron, W., Steinitz, H., Ferber, I.: The Biota of the Red Sea and the Eastern Mediterranean (1967–1972). A survey of the marine life of Israel and surroundings. Israel J. Zool. **21**(3–4), 459–523 (1972)

Por, F. D., Dor, I.: Ecology of the metahaline Pool of Di Zahav, Gulf of Elat, with notes on the Siphonocladacea and the typology of near-shore Marine Pools. Mar. Biol. **29**, 37–44 (1975a)

Por, F. D., Dor, I.: The hard bottom mangroves of Sinai, Red Sea. Rapp. Comm. Int. Explor. Mer Médit. **23**(2), 145–147 (1975b)

Por, F. D., Ferber, I.: The Hebrew University.—Smithsonian Institution collections from the Suez Canal (1967–1972). Israel J. Zool. **21**(3–4), 149–166 (1972)

Por, F. D., Marcus, A.: Copepoda Harpacticoida of the Suez Canal. Israel J. Zool. **21**(3–4), 249–274 (1972)

Potts, F. A.: XXXVI. Report on the Annelids (Sedentary Polychaets). Cambridge Expedition to the Suez Canal, 1924. Trans. Zool. Soc. London **22**, 693–705 (1928)

Powell, N. A.: Bryozoa (Polyzoa) from the South Red Sea. Cah. Biol. Mar. **8**(2), 161–183 (1967)

Powell, N. A.: Indo-Pacific Bryozoa new to the Mediterranean coast of Israel. Israel J. Zool. **18**, 157–168 (1969a)

Powell, N. A.: A checklist of Indo-Pacific Bryozoa in the Red Sea. Israel J. Zool. **18**, 357–362 (1969b)

Powell, N. A.: The marine Bryozoa near the Panama Canal. Bull. Mar. Sci. **21**(3), 766–778 (1971)

Pudney, J.: Suez: De Lesseps' Canal. London: J. M. Dent and Sons, 1968, 242 pp.

Purser, B. H., Seibold, E.: The principal environmental factors influencing holocene sedimentation and diagenesis in the Persian Gulf. In: The Persian Gulf Purser, B. H. (ed.). Berlin-Heidelberg-New York: Springer, 1973, pp. 1–9

Pusanov, I. I.: Mediterranization of the Black Sea fauna and perspectives of its intensification. Zool. Zhurn. **46**, 1277–1297 (1967) (in Russian)

Ramon, E., Friedmann, I.: The gametophyte of *Padina* in the Mediterranean. Proc. 5th Intern. Seaweed Symp. (Halifax, 1965) pp. 183–196 (1966)

Ranson, G.: Les espèces d'huitres perlières du genre *Pinctada*. Mém. Inst. Sci. Nat. Belg. (Ser. 2) **67**, 7–21 (1961)

Rayss, T.: Sur les Caulerpes de la côte Palestinienne. Palestine J. Botan. Jerusalem Ser. 2, 103–124 (1941)

Rayss, T.: Les algues tropicales de la Méditerranée orientale et leur origine probable. 8th Int. Botan. Congr. Paris, Rapp. et Comm. Sect. **17**, 148–149 (1954)

Rayss, T.: Les algues marines des côtes Palestiniennes. I. Chlorophyceae. Bull. Sea Fish. Res. Stn., Haifa (9), 1–35 (1955)
Rayss, T.: Considérations sur la flore algale de la péninsule du Sinai. Ecologie des algues marine. Colloq. Int. Centre Nat. Rech. Sci. **81**, 167–175 (1959)
Rayss, T.: Sur la présence dans la Méditerranée orientale des algues tropicales de la famille des Soliériacées. Lucr. Grădinii Botan. Acta Botan. Hort. Bucuresti 1961–1962, 91–106 (1963)
Rayss, T., Dor, I.: Nouvelle contribution à la connaissance des algues marines de la Mer Rouge. Bull. Sea Fish. Res. Stn., Haifa (34), 11–42 (1963)
Rayss, T., Edelstein, T.: Deux Caulerpes nouvelles sur les côtes Méditerranéennes d'Israël. Rev. Gén. Botan. **67**, 602–620 (1960)
Reinbold, Th.: Meeresalgen von der Insel Rhodos. Hedwigia **37**, 87–90 (1898)
Reiss, Z.: Planktonic Foraminiferids, stratotypes and a reappraisal of Neogene chronostratigraphy in Israel. Israel J. Earth Sci., **17**, 153–169 (1968)
Reiss, Z., Gvirtzman, G.: *Borelis* from Israel. Eclogae Geol. Helv. **59**(1), 437–447 (1966a)
Reiss, Z., Gvirtzman, G.: Subsurface Neogene stratigraphy of Israel. Proc. 3rd Meet. Comm. Mediterranean Neogene, Berne. Leiden: Brill, 1966b, pp. 312–364
Reiss, Z., Merling-Reiss, P., Moshkovitz, S.: Quaternary Planktonic Foraminiferida and nannoplankton from the Mediterranean continental shelf and slope of Israel. Israel J. Earth Sci. **20**, 141–177 (1971)
Remane, A., Schulz, E.: Die Strandzonen des Roten Meeres und ihre Tierwelt. Kiel. Meeresforsch. 20 (Sdbd.), 1–17 (1964)
Reyss, D.: Contribution a l'étude du rech Lacaze-Duthiers valee sous-marine des côtes de Rousillon. Vie et Milieu **15**(1), 1–46 (1964)
Riedl, R.: Fauna und Flora der Adria. Hamburg: Parey, 1963, 640 pp.
Riemann, F., Rachor, E.: Freilebende Nematoden aus dem Suezkanal. Israel J. Zool. **21**(3–4), 167–187 (1972)
*Robinet, H., Lefort, J.: Compt. Rend. Acad. Sci., Paris **62**, 436–438 (1866)
Robinson, V. C.: XXII. Report on the Sipunculoidea. Cambridge Expedition to the Suez Canal, 1924. Trans. Zool. Soc. London **22**, 359–360 (1927a)
Robinson, V. C.: XXIII. Report on a new species of Enteropneust. Cambridge Expedition to the Suez Canal, 1924. Trans. Zool. Soc. London **22**, 361–364 (1927b)
Robinson, V. C.: XXX. Report on the Nemertines. Cambridge Expedition to the Suez Canal, 1924. Trans. Zool. Soc. London **22**, 441 (1927c)
Robson, G. C.: XIX. Report on the Mollusca (Cephalopoda). Cambridge Expedition to the Suez Canal, 1924. Trans. Zool. Soc. London **22**, 321–329 (1927)
Roth, Y.: Bardawil Lagoon Project—Progress Report 1971/72. 1. Chemical Survey. Israel Oceanogr. Limnol. Res. Ltd. Haifa, 1972, pp. 1–17 (mimeo) (in Hebrew)
Roth, Y.: Salinity, Oxygen and Nutrients in the Bardawil Lagoon (Report for 1972/73). Israel Oceanogr. Limnol. Res. Ltd. Haifa, 1973, pp. 1–8 (mimeo) (in Hebrew)
Rouch, J.: Observations superficielles en Méditerranée et dans l'Atlantique. Ann. Inst. Océanogr. **20**, 51–74 (1940)
Rouch, J.: Température et salinité de la mer a Beyrouth. Bull. Inst. Océanogr. Monaco 884, 1–6 (1945)
*Roux, Ch.: L'Isthme et le Canal de Suez. Historique. État actuel. Paris. I + II, 1901
Row, R. W. H.: Report on the Sponges collected by Mr. Cyril Crossland in 1904–05. Reports on the Marine Biology of the Sudanese Red Sea. J. Linn. Soc. London (Zool.) **31**, 287–400 (1911)
Rubinoff, I.: Central American sea-level canal: possible biological effects. Science, **161**, 857–861 (1968)
Rubinoff, I.: The sea-level Canal controversy. Biological Conservation **3**(1), 33–36 (1970)
Rubinoff, R. W., Rubinoff, I.: Interoceanic colonization of a marine goby through the Panama Canal. Nature (London) **217**(5127), 476–478 (1968)
Ruggieri, G.: The Miocene and later evolution of the Mediterranean Sea. In: Aspects of Tethyan Biogeography. Adams, C. G., Ager, D. V. (eds.). The Systematics Association Publ. **7**, 283–290 (1967)

Russo, A.: Echinodermi raccolti dal Prof. L. Sanzo nella campagna della R. N. "Ammiraglio Magnaghi" in Mar Rosso e zone viciniori, 1923–1924. I–III. R. Mem. Com. Talassogr. Ital. 166; 186; 194 (1929–32)

Russo, A.: Crinoidea, Asteroidea, Ophiuroidea, Echinoidea raccolti dell R. N. "Ammiraglio Magnaghi" in Mar Rosso e zone viciniori, 1923–1924. Boll. Acc. Gioenia Catania (2), 68, 16–20 (1935)

Ryan, W. B. F.: Stratigraphy of Late Quaternary Sediments in the Eastern Mediterranean. In: The Mediterranean Sea: a Natural Sedimentation Laboratory. Stanley, D. J. (ed.). London: Hutchinson and Ross, 1972, pp. 149–169

Ryland, A. T.: The shells of the Suez Canal. Royal Air Force Education Branch, 1–27 (1951)

Safriel, U.: Vermetid gastropods and intertidal reefs in Israel and Bermuda. Science **186**, 1113–1115 (1975)

Safriel, U., Lipkin, Y.: Patterns of colonization of the Eastern Mediterranean intertidal zone by Red Sea immigrants. J. Ecol. **63**(1), 61–63 (1975)

Said, R.: The Geology of Egypt. Amsterdam, New York: Elsevier, 1962, 377 pp.

Schellenberg, A.: XXXV. Report on the Amphipoda. Cambridge Expedition to the Suez Canal, 1924. Trans. Zool. Soc. London **22**, 633–692 (1928)

Schellenberg, A.: The Fishery Grounds near Alexandria. X. Amphipoda benthonica. Notes Fish. Res. Dir. Cairo (18), 1–27 (1936)

Schleiden, M. J.: Die Landenge von Suez. Zur Beurteilung des Canalprojekts und des Auszugs der Israeliten aus Ägypten. Leipzig, 1858, 202 pp.

Schmidt, C.: Hydrologische Untersuchungen. Bull. Acad. Imp. Sci. St. Petersburg **24**, 227–235; 247–258 (1878)

Schmidt, H.: Prodromus zu einer Monographie der Mediterranen Aktinien. Zoologica **121**, 146 pp. (1972)

Schmidt, H.-E.: Some new records of hydroids from the Gulf of Aqaba with zoogeographical remarks on the Red Sea area. J. Mar. Biol. Assoc. India **13**(1), 27–51 (1971)

Schmidt, H.-E.: The distribution of Hydroida (Hydrozoa: Coelenterata) in the Red Sea and Eastern Mediterranean. 17e Congr. Intern. Zool. (Monte Carlo, Sept. 1972). Theme No. 3: Les conséquences biologiques des canaux inter-océans. 8 pp. (1972) (mimeo)

* Schmidt, J.: Danas Togt omkring Jorden 1928–1930. 1932

Schminke, H. K.: Evolution, System und Verbreitungsgeschichte der Familie Parabathynellidae (Bathynellacea, Malacostraca). Mikrofauna des Meeresbodens 24. Akad. Wissensch. Mainz (Math.-Nat. Kl.), 1–192 (1973)

Schoenfield, H. J.: The Suez Canal. London: Penguin, 1939

Schussnig, B.: Phycologische Beiträge. III. *Acetabularia Wettsteinii* n.sp. im Mittelmeer. Oestern. Botan. Zeitschr. **79**, 333–339 (1930)

* Seetzen, U. J.: Brief account of the countries adjoining the lake of Tiberias, the Jordan and the Dead Sea. London: 1813

Segerstråle, S. G.: Baltic Sea. In: Treatise on Marine Ecology and Paleoecology. Hedgpeth, J. W. (ed.). Geol. Soc. Am. Memoir **67**, 751–800 (1957)

Settepassi, F.: A proposito della *Cyprea caurica* e del *Conus mediterraneus ater*. Conchiglie, Unione Malac. Ital. Anno IV, no. 3–4, 43 (1968)

Sewell, S. R. B.: Copepoda, Harpacticoida. The John Murray Expedition 1933–1934, Sci. Rep. Brit. Mus. Nat. Hist. **7**, 117–382 (1940)

Sewell, S. R. B.: The Free-swimming planktonic Copepoda: Geographical Distribution. John Murray Exp. Sci. Rep. **8**(3), 317–595 (1948)

Simpson, G. G.: The Geography of Evolution. Philadelphia, New York: Chilton, 1965, 249 pp.

* Smith, S. A.: Remarks upon the relationship of the molluscan fauna of the Red Sea and the Mediterranean. Mediter. Naturalist 1 (1892)

Sneh, A., Weissbrod, T.: Nile Delta: The defunct Pelusiac branch identified. Science **180**, 59–61 (1973)

Sneh, A., Weissbrod, T., Perath, I.: Evidence for an Ancient Egyptian Frontier Canal. Am. Scientist **63**, 542–548 (1975)

Soar, C. D.: XVI. Report on a Hydrarachnid. Cambridge Expedition to the Suez Canal, 1924. Trans. Zool. Soc. London **22**, 287 (1927)

Spada,G.: Ritrovamenti malacologici nel mare de Lampedusa. Conchiglie, Unione Malacol. Ital. V. (1–2), 12–15 (1969)
Spada,G.: Migration de mollusques marines de la Mer Rouge à la Méditerranée. 1er Symp. C.M.A.S. (Com. Sci. Confederation Mondiale des Activités Subaquatiques), La Havana, Cuba, Sept. 1970
Steinitz,H.: On the distribution and evolution of the Cyprinodont fishes of the Mediterranean region and the Near East. Bonn. Zool. Beitr. **2**(1–2), 113–124 (1951)
Steinitz,H.: A tentative list of immigrants via the Suez Canal. Israel J. Zool. **16**(3), 166–169 (1967)
Steinitz,H.: Remarks on the Suez Canal as pathway and as habitat. Rapp. Comm. Int. Explor. Mer Médit. **19**(2), 139–141 (1968)
Steinitz,H.: A Critical List of Immigrants via the Suez Canal. In: The Hebrew University—Smithsonian Institution Joint Program: Biota of the Red Sea and the Eastern Mediterranean, Progress Report. Dept. Zool., Hebrew Univ. Jerusalem, 1970, pp.64–74 (mimeo)
Steinitz,H.: Fish ecology of the Red Sea and the Persian Gulf (Abstract). In: The Biology of the Indian Ocean, Ecological Studies, Zeitschel,B. (ed.). Berlin-Heidelberg-New York: Springer, 1973, Vol. III. pp.465–466
Steinitz,H., Ben-Tuvia,A.: Fishes of the Suez Canal. Israel J. Zool. **21**(3–4), 385–389 (1972)
Steinitz,W.: Denkschrift zur Begründung einer zoologischen Meeresstation an der Küste Palästinas. Breslau (privately printed), 1919, 24 pp.
Steinitz,W.: Beiträge zur Kenntnis der Küstenfauna Palästinas. I. Publ. Staz. Zool. Napoli **8**(3–4), 311–353 (1927)
Steinitz,W.: Die Wanderung indopazifischer Arten ins Mittelmeer seit Beginn der Quartärperiode. Intern. Rev. Hydrobiol. & Hydrogr. **22**(1–2), 1–90 (1929)
Steinitz,W.: Beiträge zur Kenntnis der Küstenfauna Palästinas. II. Publ. Staz. Zool. Napoli **13**(1), 143–154 (1933)
Steinitz,W.: Die Herkunft der Fauna in den Küstengewässern Palästinas. Soc. Natur. Sci. Tel-Aviv (privately printed) 1–7 (1936)
Stephen,A.C.: The Sipunculids of Haifa Bay and neighbourhood. Bull. Res. Coun. of Israel **7B**, 129–136 (1958)
Steuer,A.: The Fishery Grounds near Alexandria, I. Preliminary report. Notes & Mem. Fish. Res. Dir. Cairo (8), 1–18 (1935)
Steuer,A.: The Fishery Grounds near Alexandria. XVI. Cumacea, Stomatopoda and Leptostraca. Notes & Mem. Fish. Res. Dir. Cairo (26), 1–16 (1938)
Steuer,A.: The Fishery Grounds near Alexandria. XVIII. Sipunculoidea, Phoronidea, Brachiopoda, Enteropneusta and Acrania. Notes & Mem. Fish. Res. Dir., Cairo (30), 1–15 (1939a)
Steuer,A.: Die Fischereigründe vor Alexandrien. XIX Mollusca. Thalassia **3**(4), 1–143 (1939b)
Stiasny,G.: Beobachtungen über die marine Fauna des Triester Golfes. Zool. Anzeiger Bd. 32, 748 (1908)
Stirn,J.: Some notes on western trends of Lessepsian migrations. Journées ichthyol. Rome (1970), C.I.E.S.M. 187–190 (1973)
Stock,J.H.: Pycnogonida from the Mediterranean coast of Israel. Bull. Res. Coun. Israel **7B**(3–4), 137–142 (1958)
Stoffers,P., Kuhn,R.: Red Sea evaporites: A petrographic and geochemical study. In: Initial Reports of DSDP 23. Whitmarsh,R.B., Weser,O.E., Ross,D.A. (eds.). U.S. Govt. Printing Press, 1974, pp. 821–847
Stoffers,P., Ross,D.A.: Sedimentary history of the Red Sea. In: Initial Reports of DSDP 23. Whitmarsh,R.B., Weser,O.E., Ross,D.A. (eds.). U.S. Govt. Printing Press, 1974, pp.849–865
Sturany,R.: Lamellibranchiaten des Rothen Meeres. Expeditionen S.M. Schiff „Pola" in das Rothe Meer (1895/96 und 1897/98). Zool. Ergeb. Wien **14**, 1–41 (1899)
Sturany,R.: Gastropoden des Rothen Meeres. Expeditionen S.M. Schiff „Pola" in das Rothe Meer (1895/96 und 1897/98). Zool. Ergeb. Wien **23**, 1–75 (1903)
*Svedelius,N.: On the discontinuous geographical distribution of some tropical and subtropical marine algae, Ark. Botan. **19**(3) (1924)

Swartz, D. H., Arden, D. D. Jr.: Geologic History of the Red Sea Area. Bull. Am. Assoc. Petrol. Geol. **44**(10), 1621–1637 (1960)

Swennen, C.: On a collection of Opisthobranchia from Turkey. Zool. Meded. Rijksmus. Nat. Hist. Leiden **38**(3), 41–75 (1961)

Talmadge, R. R.: Notes on Israeli Haliotids. Argamon, J. Israel Malacol. Soc. **2**, 81–85 (1971)

Tardent, P.: Capture d'un *Abudefduf saxatilis vaigiensis* Q. et G. (Pisces, Pomacentridae) dans le Golfe de Naples. Rev. Suisse Zool. **66**(20), 347–351 (1959)

Tattersall, W. M.: XI. Report on the Crustacea Mysidacea. Cambridge Expedition to the Suez Canal, 1924. Trans. Zool. Soc. London **22**, 185–199 (1927)

Taylor, F. J. R.: General features of dinoflagellate material collected by the "Anton Bruun" during the International Indian Ocean Expedition. In: The Biology of the Indian Ocean. Zeitschel, B. (ed.). Berlin-Heidelberg-New York: Springer, 1973, pp. 155–170

*Tchukhtchin, V. D.: Trudy Sevastopol Biol. Stn. 16, 215–223 (1964) (in Russian)

Tebble, N.: On a collection of Polychaetes from the Mediterranean coast of Israel. Bull. Res. Coun. Israel **8B**(1), 9–30 (1959)

Tendler, A.: Studies on the biology of the japanese shrimp *Penaeus japonicus* Bate (Crustacea, Decapoda) in the Nahal Rubin area off the Mediterranean coast of Israel. Master's thesis, Hebrew Univ. Jerusalem, 1972 (in Hebrew)

Thompson, I. C., Scott, A.: Report on the Copepoda collected by Professor Herdman at Ceylon in 1902. Ceylon Pearl Oyster Fish. Suppl. Report **7**, 227–307, 20 plates (1903)

Thomsen, H.: Nitrate and phosphate contents of the Mediterranean water. Rep. Danish Oceanogr. Exped. Medit. 1908–1910, **3**(6), 1–4 (1931)

Thornley, L. R.: Reports on the Marine Biology of the Sudanese Red Sea. X. Hydroida. J. Linn. Soc. London (Zool.) **31**, 80–85 (1908)

Thorpe, W. H.: XXXII. Report on the brackish water insects. Cambridge Expedition to the Suez Canal, 1924. Trans. Zool. Soc. London **22**, 447–449 (1927)

Thorson, G.: Animal migrations through the Suez Canal in the past, recent years and the future (a preliminary report). Vie et Milieu. Suppl. (22), 841–846 (1971)

Tillier, J. B.: Le Canal de Suez et sa faune ichthyologique. Mém. Soc. Zool. France **15**(3–4), 279–318 (1902)

Tillier, L., Bavay, A.: Les mollusques testacés du Canal de Suez. Bull. Soc. Zool. France **30**, 170–181 (1905)

Tillier, L., Bavay, A.: Au sujet des mollusques testacés du Canal de Suez. Bull. Soc. Zool. France **31**, 129–131 (1906)

Tom, M.: The Benthic Fauna Association of Haifa Bay. M. Sc. Thesis, Tel-Aviv University (in Hebrew) (1976)

Tomlin, J. R. (le B.): XVIII. Report on the Mollusca (Amphineura, Gastropoda, Scaphopoda, Pelecypoda). Cambridge Expedition to the Suez Canal, 1924. Trans. Zool. Soc. London **22**, 291–320 (1927)

Topp, R. W.: Interoceanic sea-level canal: effects on the fish faunas. Science **165**(3900), 1324–1327 (1969)

Tortonese, E.: Ricerche Zoologiche nel Canale de Suez e dintorni. I. Echinodermi. Rendiconti dell'Accademia Nazionale de Lincei (Classe di Scienze Fisiche, Matematiche e Naturali), Serie VIII, **2**(6), 835–838 (1947)

Tortonese, E.: Ricerche Zoologiche nel Canale di Suez e dintorni. II. Pesci. Arch. Zool. Ital., **33**, 275–292 (1948)

Tortonese, E.: I caratteri biologici del Mediterraneo orientale e i problemi relativi. Attualita Zoologiche Torino Arch. Zool. Ital. Suppl. **7**, 205–251 (1951)

Tortonese, E.: Some field-notes on the fauna of the Suez Canal (Timsah and Bitter Lakes). Hydrobiologi Istanbul (Ser. B) **1**(1), 1–6 (1952)

Tortonese, E.: Gli Echinodermi viventi presso le coste dello stato di Israele. (Mar di Levante, Golfo di Elath). Boll. Ist. Zool. Univ. Torino **4**(4), 1–35 (1953–54)

Tortonese, E.: The relations between the Mediterranean and Atlantic Fauna. Hidrobiologi, Istanbul (Ser. B) **5**(1–2), 30–34 (1960)

Tortonese, E.: The main biogeographical features and problems of the Mediterranean fish fauna. Copeia 1964(1), 98–107 (1964)

Tortonese, E.: Echinoderms from the Coast of Lebanon. Misc. Papers Nat. Sci. Am. Univ. Beirut **5**, 2–5 (1966)

Tortonese, E.: La Fauna del Mediterraneo e i suoi rapporti con quelle dei mari vicini. Pubbl. Staz. Zool. Napoli **37**, Suppl. 2, 369–384 (1969)

Tortonese, E.: On the occurrence of *Siganus* (Pisces) along the coast of North Africa. Doriana, **4**(191), 1–2 (1970)

Tortonese, E.: Rapports entre la faune méditerranéenne de poissons percoïdes et celles des mers voisines (Atlantique oriental, mer Noire et mer Rouge). Journées ichthyol. Rome (1970) C.I.E.S.M., 39–42 (1973a)

Tortonese, E.: Facts and perspectives related to the spreading of Red Sea organisms into the Eastern Mediterranean. Ann. Mus. Civico Storia Nat. Genova **79**, 322–329 (1973b)

Tsurnamal, M.: Sponges of Red Sea origin on the Mediterranean coast of Israel. Israel J. Zool. **18**(2–3), 149–155 (1969a)

Tsurnamal, M.: Four new species of Mediterranean Demospongiae and new data on *Callites lacazii* Schmidt. Cah. Biol. Mar. **10**(4), 343–357 (1969b)

Udvardy, M.D.F.: Dynamic Zoogeography with special reference to Land Animals. New York: Van Nostrand Reinhold, 1969, 445 pp.

Vaillant, L.: Recherches sur la faune malacologique de la baie de Suez. J. Conchyl. Paris **13**, 97–127 (1865)

*Vassel, E.: Sur la Communication possible des mers d'Europe, Indo-Pacifique et des Antilles durant la periode miocene. (1889)

*Vassel, E.: Sur les faunes de l'Isthme de Suez. Bull. Soc. Hist. Natur. Autun 3, 15–99 (1890)

*Vassel, E.: Sur la Pintadine du Golfe de Gabès. Ass. Fr. Advan. Sci. Carthage (1896)

Vatova, A.: The Fishery Grounds near Alexandria II. A bottom sample taken at Alexandria. Notes Mem. Fish. Res. Dir. Cairo (9), 1–3 (1935)

*Vercelli, F.: Osservazioni Exequite nel Canale di Suez. Campagna Idrografica nel Mar Rosso della Regia Nave "Ammiraglio Magnaghi" 1923–1924 (1927)

*Victor, St.: Le Canal de Suez. Paris. 1934

*Voisin, Bey.: Le Canal de Suez. Paris 1902–1906

Voss, G.L.: Zoogeographical considerations of the impact of a sea-level canal across Panama. 17e Congr. Intern. Zool. (Monte Carlo, Sept. 1972). Theme No. 3: Les consequences biologiques des canaux inter-océans (1972) 20 pp. (mimeo)

Wainwright, S.A.: Reef Communities visited by the Israel South Red Sea Expedition, 1962. Bull. Sea Fish. Res. Stn., Haifa (38), 40–53 (1965)

Wells, J.B.J.: The Littoral Copepoda (Crustacea) of Inhaca Island, Mozambique. Trans. R. Soc. Edinburgh **67**(7), 189–358 (1967)

Wesenberg-Lund, E.: Sipunculoidea from the coast of Israel. Bull. Res. Coun. of Israel **6B**, 193–200 (1957a)

Wesenberg-Lund, E.: Sipunculoidea and Echiuroidea from the Red Sea. Bull. Sea Fish. Res. Stn., Haifa (14), 1–15 (1957b)

Wetzel, R., Haller, J.: Sur le Quaternaire côtier de la region de Tripoli (Liban). Notes Mem. Sect. Géol. Beyrouth **4**, 1–48 (1945)

Whitehead, P.J.P.: A revision of the recent round herrings (Pisces: Dussumieriidae). Bull. Brit. Mus. Nat. Hist. (Zool.) **10**(6), 305–380 (1963)

Whitmarsh, R.B.: Summary of general features of Arabian Sea and Red Sea Cenozoic, based on Leg. 23 cores. In: Initial Reports of DSDP 23. Whitmarsh, R.B., Weser, O.E., Ross, D.A. (eds.). U.S. Govt. Printing Press, 1974, pp. 1115–1123

Wimpenny, R.S.: Some hydrographic data from the Suez Canal 1928–1929. Bull. Coastguards Fish. Ser. Egypt. (1), 1–7 (1930)

Wimpenny, R.S.: *Mugil seheli* (Forsk.). Faune et Flore de la Méditerranée. Rapp. Comm. Intern. Explor. Mer Médit. (1931)

Wirszubski, A.: On the biology and biotope of the red mullet, *Mullus barbatus* L. Bull. Sea Fish. Res. Stn., Caesarea (7), 1–20 + XXXII (1953)

Wolstedt, P.: Grundlinien einer Geologie des Quartärs. Das Eiszeitalter. Die Mittelmeerländer IV. pp. 368–376 (1958)

Wüst, G.: Salzgehalt und Wasserbewegung im Suezkanal. Naturwissenschaften **22**, 446–450 (1934)

Wüst, G.: Fortschreitende Salzgehaltsabnahme im Suezkanal. Annal. Hydrogr. Marit. Meteorol. Berlin **63**(10), 391–395 (1935)
Wüst, G.: Über die Abnahme des Salzgehalts im Suezkanal von 1869 bis 1937. Erdkunde **5**, 241–243 (1951)
Zaccaria, M.: Studi geologici sulle isole del Dodecaneso (Mare Egeo); VIII. Fauna Calabriana dell'Isola di Rodi. Riv. Ital. Paleont. **74**(1), 275–306 (1968)
Zanardini, J.: Plantarum in mari rubro hucusque collectarum enumeratio. Mem. Ist. Veneto Sci. **7**, 209–309 (1858)
Zenkevitch, L.: Biology of the Seas of the USSR. London: Allen and Unwin, 1963, 955 pp.
Zeuner, F. E.: The Pleistocene Period, its Climate, Chronology, and Faunal Succession. London: Hutchinson, 1959, 447 pp.
Zibrowius, H.: Serpulidae (Ann. Polych.) des côtes ouest de l'Afrique et des archipels voisins. Annales Musée Trevuren Sciences Zool. 207, 93 pp. (1973)

Plates

1
2
3
4
6
5

Plate 2. Lessepsian migrant Mollusca, Gastropoda. *(1) Rissoina bertholetti; (2) Cerithium scabridum; (3) Cerithium kochi; (4) Thais carinifera; (5) Murex tribulus.* (*1–4* from Barash and Danin, 1972; *5* original)

◄ Plate 1. Lessepsian migrant Mollusca, Lamellibranchia (Pelecypoda) *(1) Pinctada radiata; (2) Malleus regula; (3) Brachidontes variabilis; (4) Paphia textile; (5) Clementia papyracea; (6) Mactra olorina* (*1* original; *2–6* from Barash and Danin, 1972)

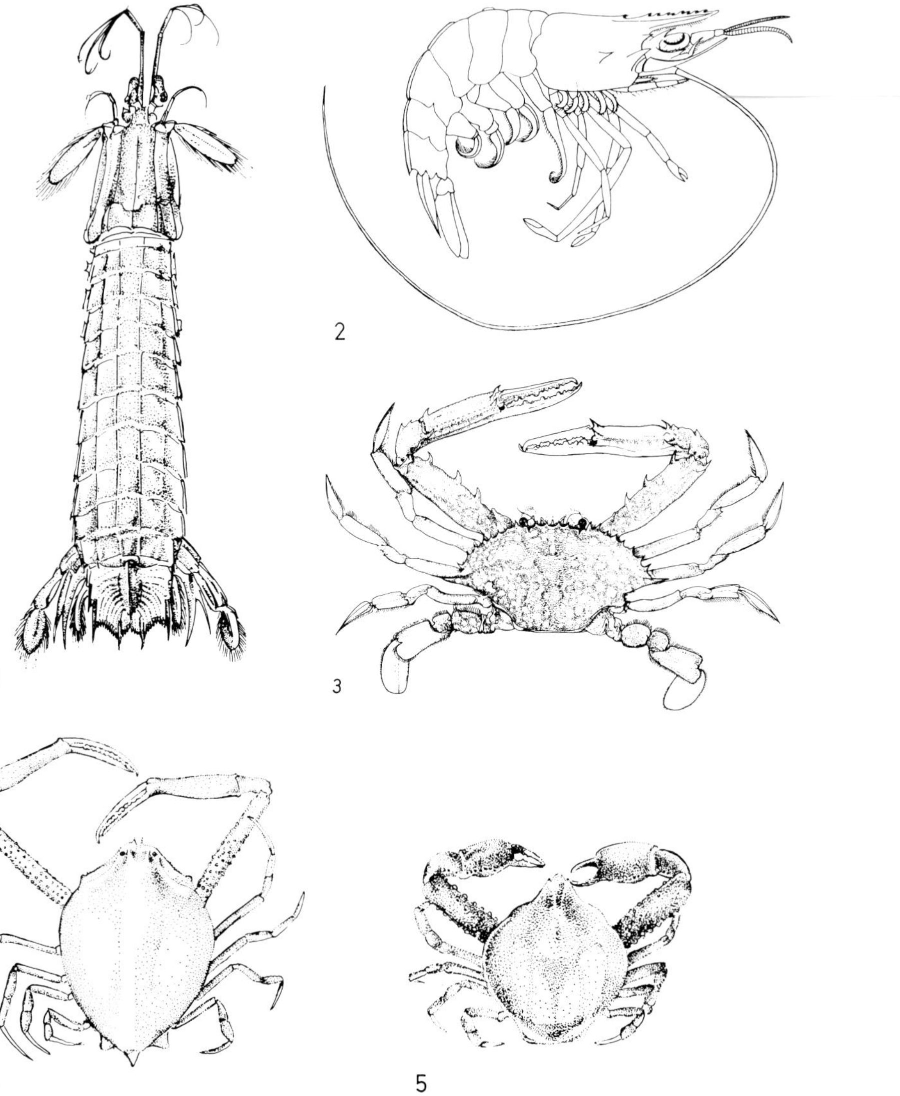

Plate 3. Lessepsian migrant Crustacea. *(1) Squilla massawensis* (Stomatopoda) (from Por, 1971b); *(2) Trachypenaeus curvirostris* (Decapoda, Macrura) (from W. Steinitz, 1927); *(3) Portunus pelagicus* (Decapoda, Brachyura) (original); *(4) Myra fugax* (Decapoda, Brachyura) (from Por, 1971b); *(5) Leucosia signata* (Decapoda, Brachyura) (from Por, 1971b)

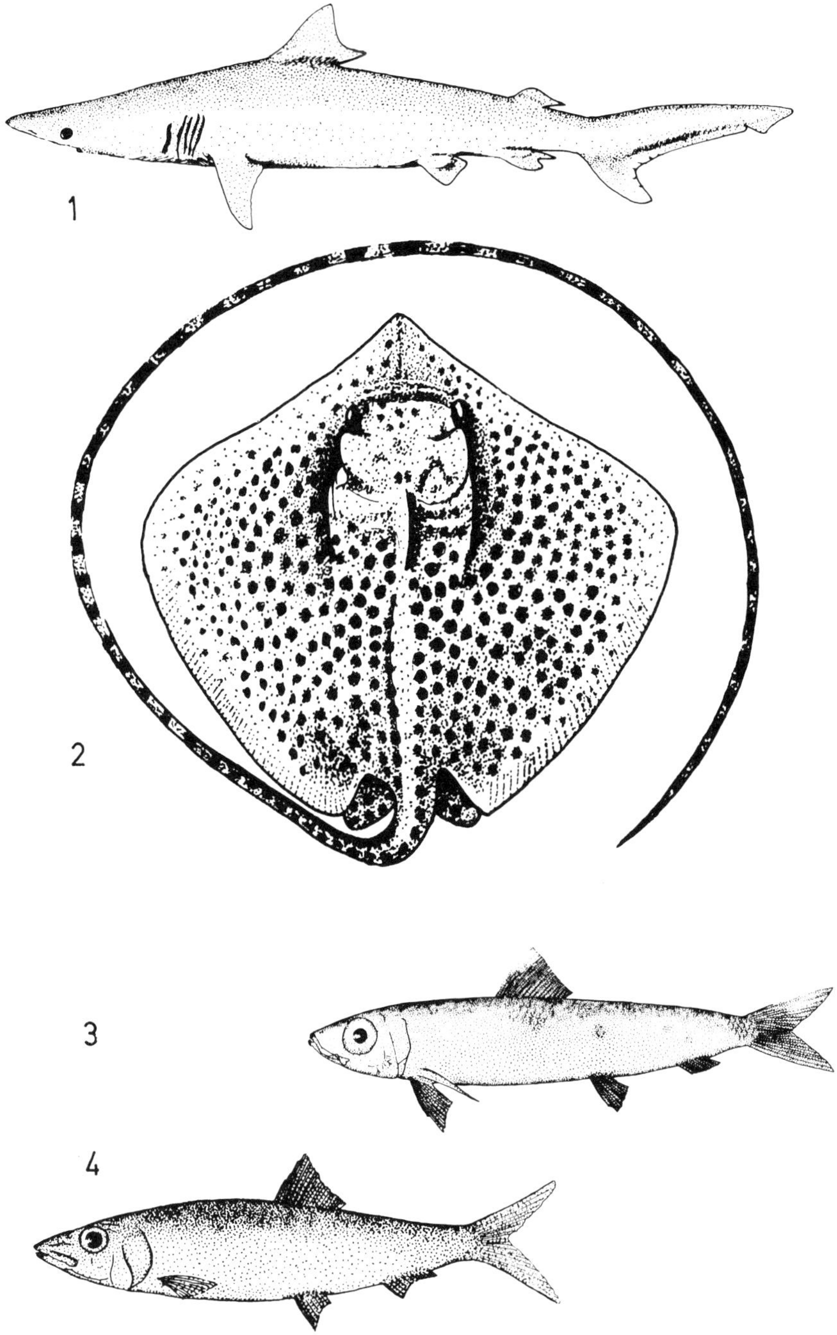

Plates 4–10. Lessepsian migrant fishes (from Ben-Tuvia, 1976)

Plate 4. *(1) Carcharhinus brevipinna; (2) Himantura uarnak; (3) Etrumeus teres; (4) Dussumieria acuta*

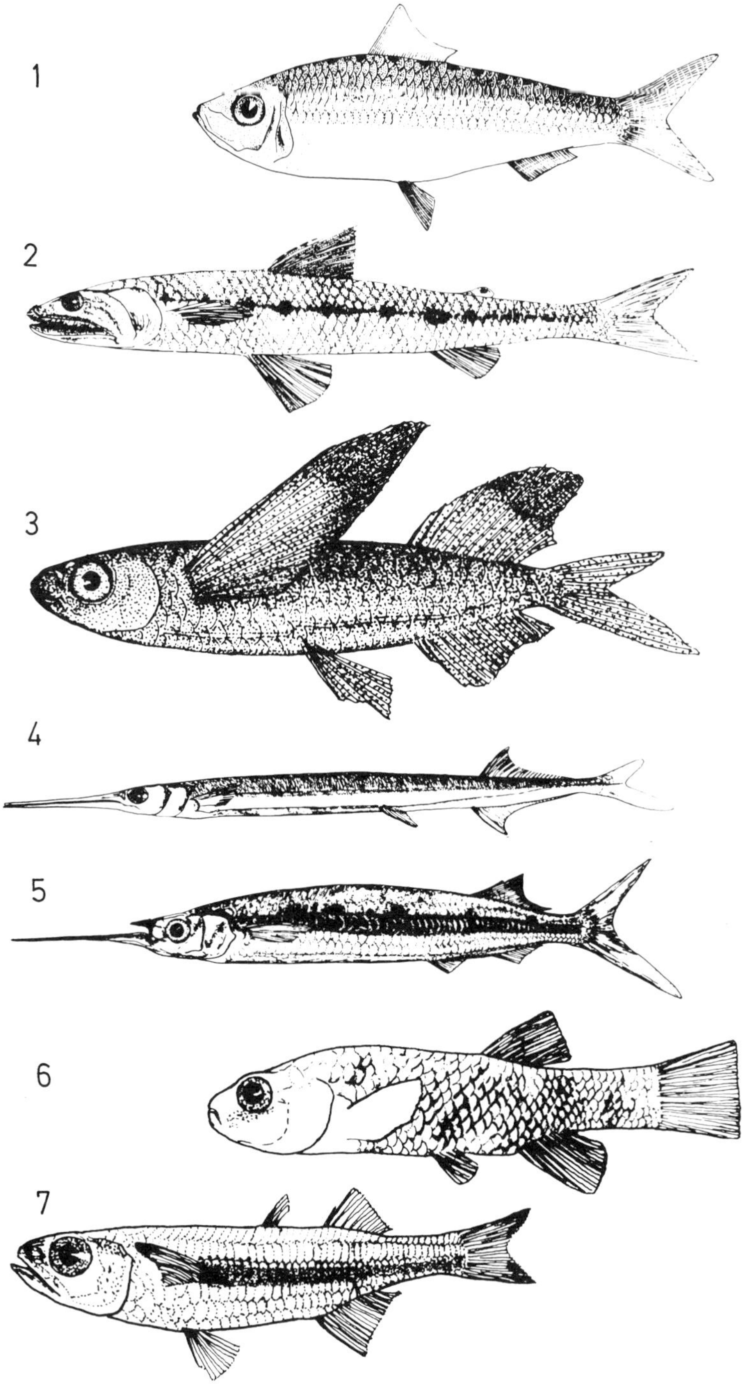

Plate 5. *(1)* **Herclotsichthys** *punctatus; (2) Saurida undosquamis; (3) Parexocoetus mento; (4) Tylosurus choram; (5) Hemiramphus far; (6) Aphanius dispar; (7) Pranesus pinguis*

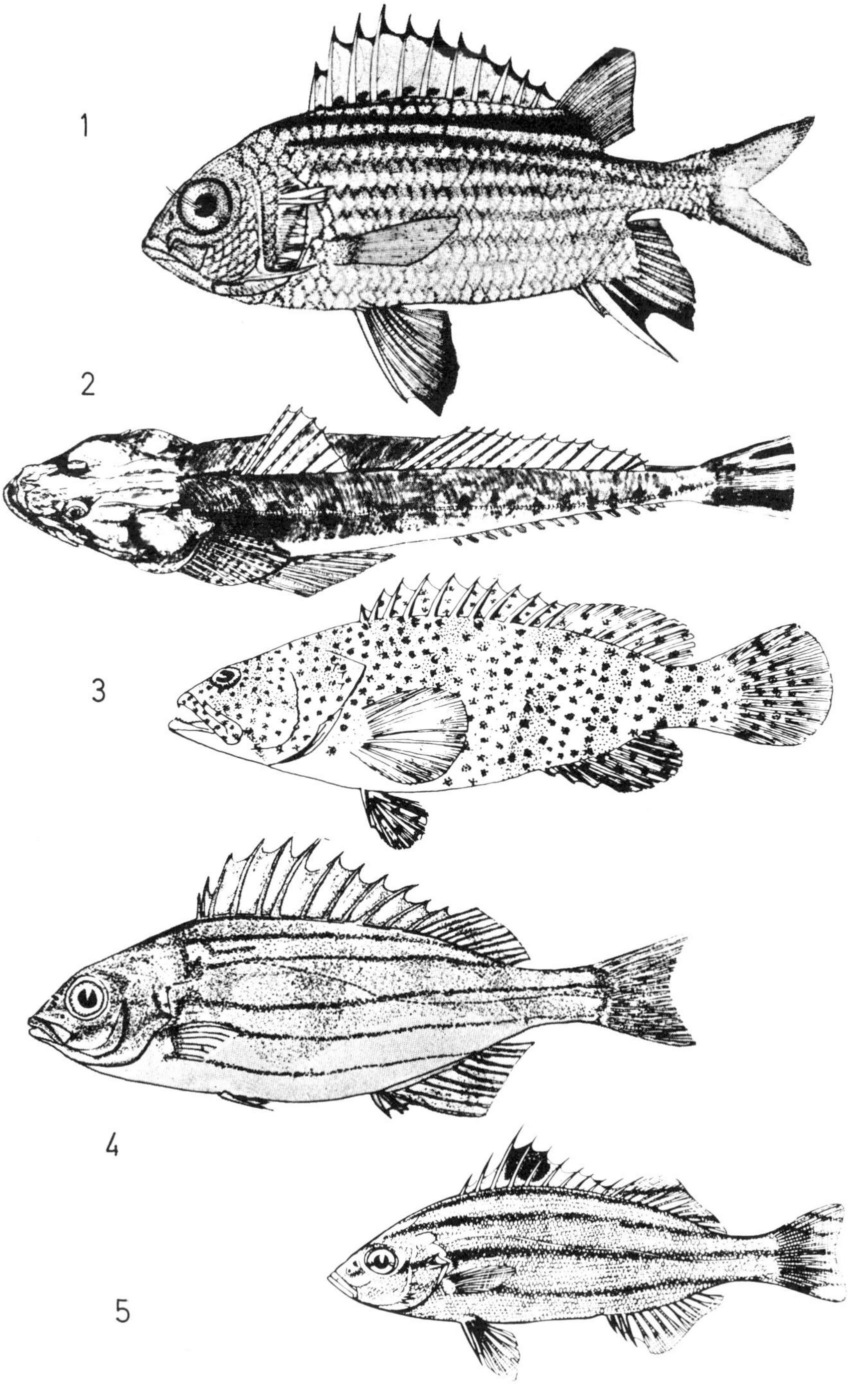

Plate 6. *(1) Holocentrus ruber; (2) Platycephalus indicus; (3) Epinephelus tauvina; (4) Pelates quadrilineatus; (5) Therapon puta*

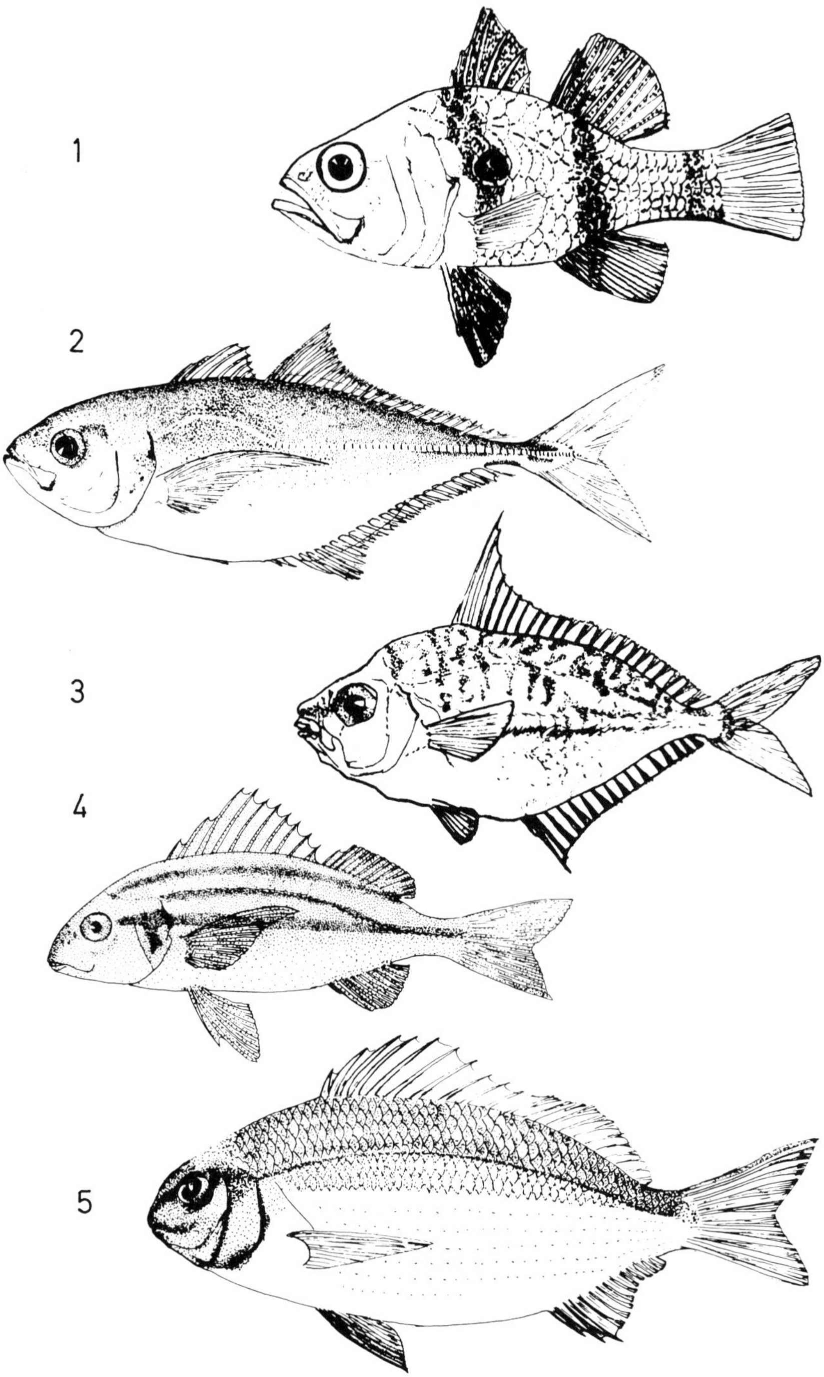

Plate 7. *(1) Apogonichthyoides nigripinnis; (2) Atule djeddaba; (3) Leiognathus klunzingeri; (4) Rhonciscus stridens; (5) Crenidens crenidens*

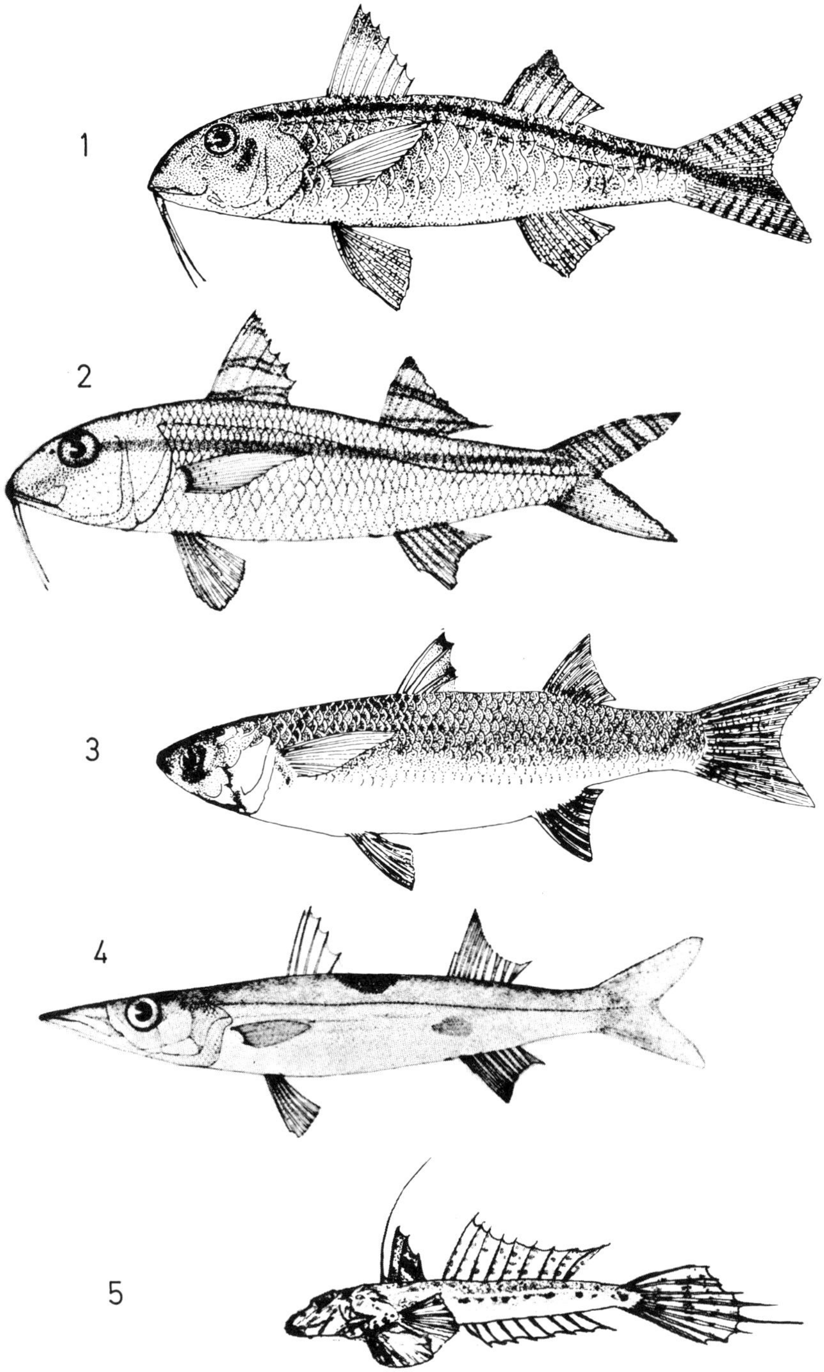

Plate 8. *(1) Upeneus asymmetricus; (2) Upeneus moluccensis; (3) Liza carinata; (4) Sphyraena chrysotaenia; (5) Callionymus filamentosus*

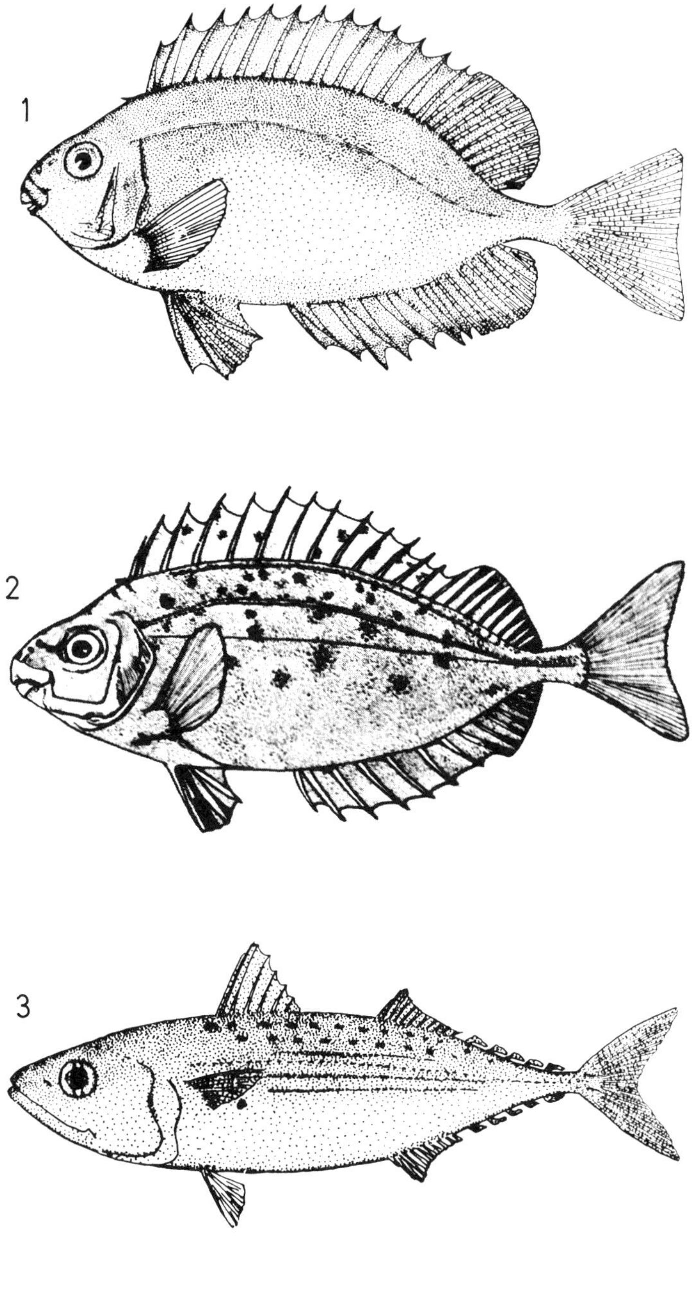

Plate 9. *(1) Siganus luridus; (2) Siganus rivulatus; (3) Rastrelliger kanagurta*

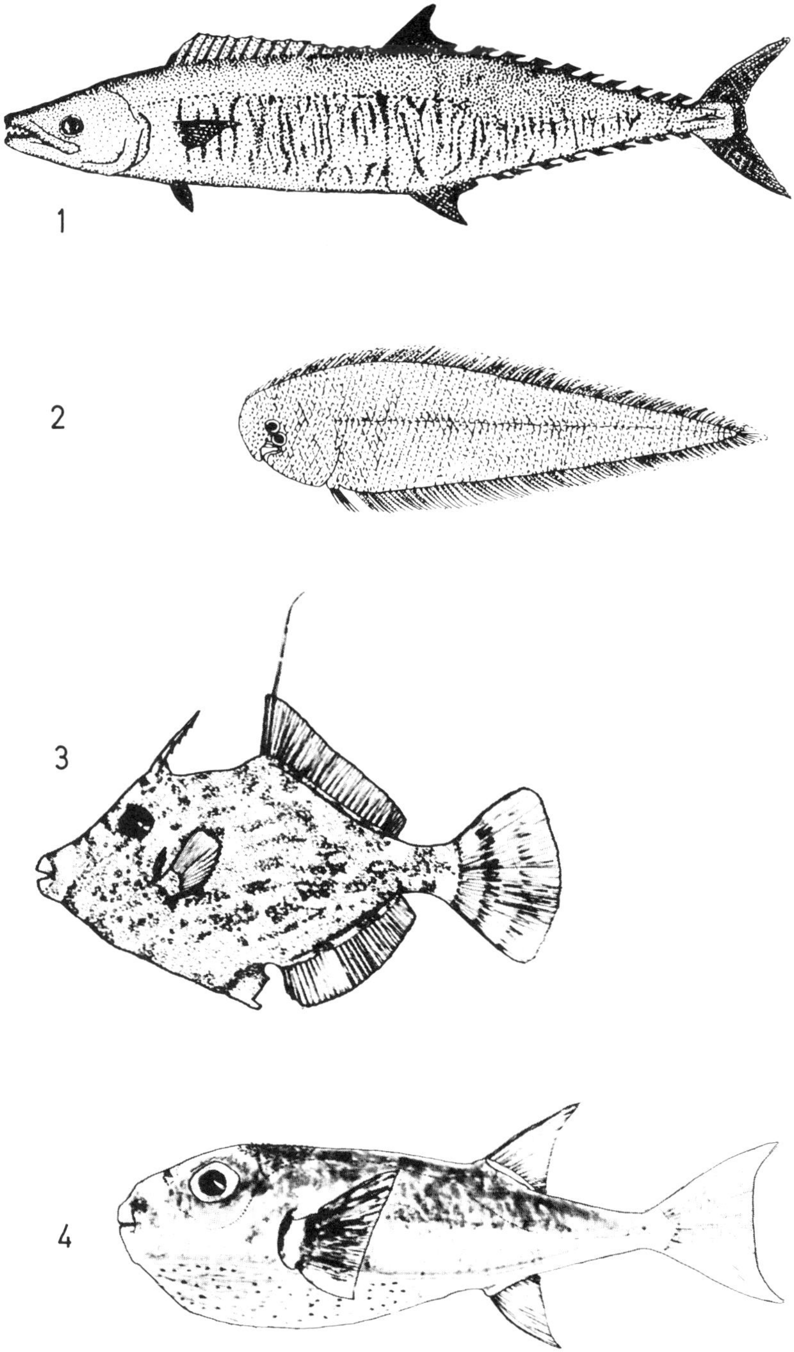

Plate 10. *(1) Scomberomorus commerson; (2) Dollfusichthys sinusarabici; (3) Stephanolepis diaspros; (4) Sphaeroides spadiceus*

Taxonomic Index*

* Compiled by Miss I. Ferber.

Subject and Geographical Index*

Page numbers in bold face refer to main entries (chapters).

* *Note*: Geographical names mentioned in the text are included. Due to discrepancies in the spelling of names transliterated from Hebrew and Arabic, alternative spelling may be given in parentheses. Spelling of localities in Israel and Sinai is according to: "Israel — Map of the Cease-fire Lines 1967" (Survey of Israel). See maps in text pp. 43, 85.
Compiled by I. Ferber.

Ecological Studies

Analysis and Synthesis

Editors: W.D. Billings, F. Golley, O.L. Lange, J.S. Olson

Volume 1

Analysis of Temperate Forest Ecosystems

Editor: D.E. Reichle
1st corrected reprint
91 figures. XII, 304 pages. 1973
ISBN 3-540-04793-X

Volume 2

Integrated Experimental Ecology

Methods and Results of Ecosystem Research in the German Solling Project
Editor: H. Ellenberg
(out of print)

Volume 3

The Biology of the Indian Ocean

Editor: B. Zeitzschel in cooperation with S.A. Gerlach
286 figures. XIII, 549 pages. 1973
ISBN 3-540-06004-9

Volume 4

Physical Aspects of Soil Water and Salts in Ecosystems

Editors: A. Hadas, D. Swartzendruber, P.E. Rijtema, M. Fuchs, B. Yaron
221 figures, 61 tables. XVI, 460 pages. 1973
ISBN 3-540-06109-6

Volume 5

Arid Zone Irrigation

Editors: B. Yaron, E. Danfors, Y. Vaadia
181 figures. X, 434 pages. 1973
ISBN 3-540-06206-8

Volume 6
K. Stern, L. Roche

Genetics of Forest Ecosystems

70 figures. X, 330 pages. 1974
ISBN 3-540-06095-2

Volume 7

Mediterranean Type Ecosystems

Origin and Structure

Editors: F. di Castri, H.A. Mooney
88 figures. XII, 405 pages. 1973
ISBN 3-540-06106-1

Volume 8

Phenology and Seasonality Modeling

Editor: H. Lieth
120 figures, XV, 444 pages. 1974
ISBN 3-540-06524-5

Volume 9
B. Slavik

Methods of Studying Plant Water Relations

With contributions by B. Slavík, J. Čatský, J. Solárová, H.R. Oppenheimer, J. Hrbáček, J. Slavíková, V. Kozinka, J. Úlehla, P.G. Jarvis, M.S. Jarvis
181 figures, XVIII, 449 pages, 1974
ISBN 3-540-06686-1

Volume 10

Coupling of Land and Water Systems

Editor: A.D. Hasler
95 figures. XVII, 309 pages. 1975
ISBN 3-540-06707-8

Volume 11

Tropical Ecological Systems

Trends in Terrestrial and Aquatic Research
Editors: F.B. Golley, E. Medina
131 figures. XV, 398 pages. 1975
ISBN 3-540-06706-X

Volume 12

Perspectives of Biophysical Ecology

Editors: D.M. Gates, R.B. Schmerl
215 figures. XIII, 609 pages. 1975
ISBN 3-540-06743-4

Volume 13

Epidemics of Plant Diseases

Mathematical Analysis and Modeling
Editor: J. Kranz
46 figures. X, 170 pages. 1974
ISBN 3-540-06896-1

Volume 14

Primary Productivity of the Biosphere

Editors: H. Lieth, R.H. Whittaker
With contributions by E. Box, J.S. Bunt, C.A.S. Hall, H. Lieth, G.E. Likens, P.L. Marks, R. Moll, P.G. Murphy, D.D. Sharp, D. Sharpe, D. Whigham, R.H. Whittaker
67 figures, 46 tables. VIII, 339 pages. 1975
ISBN 3-540-07083-4

Volume 15

Plants in Saline Environments

Editors: A. Poljakoff-Mayber, J. Gale
With contributions by: D.L. Carter, V.J. Chapman, L.D. Doneen, J. Gale, A. Kylin, A.J. Peck, A. Poljakoff-Mayber, R.S. Quatrano, I. Shainberg, W.W. Thomson
54 figures. VII, 213 pages. 1975
ISBN 3-540-07193-8

Volume 16 + 17

Fennoscandian Tundra Ecosystems (In 2 Parts)

Part 1 Plants and Microorganisms
Editor: F.E. Wielgolaski
Editorial Board: P. Kallio, T. Rosswall
90 figures, 96 tables. XV, 366 pages. 1975
ISBN 3-540-07218-7
Part 2: Animals and Systems Analysis
Editor: F.E. Wielgolaski
Editorial Board: P. Kallio, H. Kauri, E. Østbye, T. Rosswall
81 figures, 97 tables. XIII, 337 pages. 1975
ISBN 3-540-07551-8

Volume 18

Remote Sensing for Environmental Sciences

Editor: E. Schanda
With contributions by R.T.H. Collis, D.J. Creasey, R.L. Crasty, P. Hartl, G.P. de Loor, P.B. Russel, A.E. Salerno, E. Schanda, P.W. Schaper
178 figures (7 color plates), 31 tables. XIII, 367 pages. 1976
ISBN 3-540-07465-1

Volume 19

Water and Plant Life

Problems and Modern Approaches
Editors: O.L. Lange, L. Kappen, E.-D. Schulze
178 figures, 66 tables. XX, 536 pages. 1976
ISBN 3-540-07838-X

Volume 20
F.B. Christiansen, T.M. Fenchel

Theories of Populations in Biological Communities

68 figures, 4 tables. X, 144 pages. 1977
ISBN 3-540-08010-4

Volume 21
V.Y. Alexandrov

Cells, Molecules, and Temperature

Conformational Flexibility of Macromolecules and Ecological Adaptation
Translated from the Russian by V.A. Bernstam
74 figures, 30 tables. XII, 332 pages. 1977
ISBN 3-540-08026-0

Volume 22
R. Guderian

Air Pollution

Phytotoxicity of Acidic Gases and Its Significance in air Pollution Control
Translated from the German by C.J. Brandt
40 figures, 4 in color, 26 tables. VIII, 127 pages. 1977
ISBN 3-540-08030-9

Springer-Verlag
Berlin Heidelberg New York